I0047955

William A. (William Adolph) Baillie-Grohman, William ... ..
Baillie-Grohman

Tyrol And The Tyrolese

The People And The Land

William A. (William Adolph) Baillie-Grohman, William Adolph Baillie-Grohman

Tyrol And The Tyrolese
The People And The Land

ISBN/EAN: 9783741149306

Manufactured in Europe, USA, Canada, Australia, Japa

Cover: Foto ©Klaus-Uwe Gerhardt /pixelio.de

Manufactured and distributed by brebook publishing software
(www.brebook.com)

William A. (William Adolph) Baillie-Grohman, William Adolph
Baillie-Grohman

Tyrol And The Tyrolese

TYROL AND THE TYROLESE:

THE PEOPLE AND THE LAND

IN THEIR

SOCIAL, SPORTING, AND MOUNTAINEERING ASPECTS.

BY

W. A. BAILLIE GROHMAN.

THE CHAMOIS' HOME

WITH NUMEROUS ILLUSTRATIONS.

LONDON:

LONGMANS, GREEN, AND CO.

1876.

All rights reserved.

PREFACE.

A CERTAIN VALUE may, I hope, be imparted to this volume by the fact that I have lived for many years in the Tyrol, and being by parentage half an Austrian, and as well acquainted with the German language as with my mother tongue, am therefore more likely to gain a true insight into the lives and characters of the Tyrolese than most writers on the same subject, who have not this advantage.

My love for sport and a sound bodily constitution have gone hand in hand in enabling me to acquire an accurate acquaintance with the rough fashions of this picturesque country; and while they have brought me across many an odd character lost to the world in some out-of-the-way nook among these little-known mountains and valleys, I have had many adventures, some of which I have endeavoured to relate in the following pages.

My sketches will perhaps serve to bring before my reader's eyes the various scenes they represent in as vivid a light as I could wish.

It seems that some question has been raised relative to the spelling of the word Tyrol. Without wishing to enter more fully into the merits of the controversy, I may mention that Tyrol was up to the beginning of this century, with hardly any exception, spelt with a 'y.' It is only within the last fifty or sixty years that the letter 'i' has supplanted it, and at present we find that the word is generally spelt Tirol. The fact that a number of geographical names have undergone in this half century precisely the same change as the word Tyrol, and that the 'foreign' letter 'y' is hardly ever used by Germans, does not render the spelling of the word Tirol less incorrect, for we must remember throughout this whole question that the derivation of Tyrol is not, as many suppose, from 'Terioles,' but from 'Tyr,' a 'fortress in the mountains,' in which sense we find it in use as early as the ninth century.

I may finally remark that two of the chapters in this volume have appeared in the shape of sketches in the 'Alpine Journal.'

Schloss Matzen, Brixlegg, Tyrol :

December, 1875.

CONTENTS.

—◦◦◦·—

CHAPTER II.

PRIESTHOOD AND SUPERSTITION.

CHAPTER III.

A PEASANT'S WEDDING.

CHAPTER IV.

THE WOODCUTTER.

CHAPTER V.

THE CHAMOIS AND THE CHAMOIS STALKER.

CHAPTER VI.

THE GOLDEN EAGLE AND ITS EYRIE.

CHAPTER VII.

AN ENCOUNTER WITH TYROLESE POACHERS.

CHAPTER VIII.

A TYROLESE SMUGGLER AND HIS LIFE.

CHAPTER IX.

THE BLACKCOCK.

CHAPTER X.

A WINTER ASCENT OF THE GROSS GLOCKNER.

LIST OF ILLUSTRATIONS.

Erratum.

Page 6, line 8, *for* disastrous to them *read* disastrous to it.

MALE AND FEMALE 'WILDHEUER.'

TYROL AND THE TYROLESE.

CHAPTER I.

A GLIMPSE AT THE LANDSCAPE AND THE PEOPLE.

IT may well amaze even those who have been whirled in the train through the two or three chief valleys of Tyrol, to learn that this country, with a population considerably less than half that of Yorkshire, contains five hundred and thirty-seven old castles.

These Tyrolese castles form so picturesque a feature in scenery nearly always grand and striking, that the indulgent reader will excuse my inviting him to visit one of their number ere I lay before him the results of my experience amongst the people. To this end he will kindly accompany me up the steep path leading to the ponderous iron-

B

barred old gate giving entrance to one of the most ancient and historically interesting of Tyrolese castles—the home of this volume,—and after ascending endless flights of stairs, find himself comfortably seated in an armchair in front of the broad old-fashioned window overlooking the whole of the country near.

Lying at your feet is a goodly stretch of the smiling, exquisitely verdant valley of the Inn, skirted by two parallel rows of noble peaks terminating in the far distance with the glistening glacier world of the Oetz and Stubai Thäler.

As your eye glances down the giddy height and follows the upward course of the broad swift Inn at your feet, as it winds like a band of silver through green meadows, eight old castles, the remains of what were once feudal strongholds, occupying the eminences of hills, or perched like swallows' nests on the precipitous slopes of the adjacent mountains, become discernible. Interspersed between these hoary relics rise the amazingly slender, needle-shaped spires of three churches, the houses belonging to each village clustering round the sacred edifice. Of the broad-roofed houses, hidden behind groves of apple or nut-trees, little is to be seen ; and of such as are visible, the greater part are of the velvety brown timber which is so sunny and pleasing to the eye. Only the blue rings of smoke curling up in the gloriously-tinted evening sky indicate the presence of human habitations secreted behind bowers of trees. Fancy a dark green background of precipitously rising mountains, covered with sombre pine forest, terminating in the gray cliffs that form the eminences, thereby bringing the rich vegetation of the verdant valley into close contrast with the sternness of the impending peaks, and you have the type of a peaceful sunny North Tyrolese landscape.

I say North Tyrolese, for Tyrol, divided into halves by the high snow-peaked main chain of the Alps, represents, taken as a whole, two geographically distinct countries. North Tyrol can be identified to all practical purposes with the German cantons of Switzerland, having an Alpine climate, while the South, with its vineyards and its genial air, is akin to fertile Italy. This perfect dissimilarity of Northern to Southern Tyrol renders a cursory glance at the physical appearance of the latter indispensable in order to form a faithful conception of the whole country.

Removing our chair of observation to a window of any one of the numerous castles of Meran in South

Tyrol, we have, though at a distance of scarcely more than seventy-five miles, as the crow flies, from our former point of view, a landscape before our eyes as different from the first as it well can be.

To the painter's palette supplied with various shades of green and gray sufficient to depict North Tyrolese scenery, we have to add the blue, yellow, and mauve of Italian landscape.

The number of castles in our picture has increased from eight to five and twenty or thirty. The rich verdant pasturages are supplanted either by scrubby brushwood scorched to a sombre brown, or by large expanses of vineyards, while the dark green peaceful pine forests have been replaced by the stunted fir of a brownish tint, or by the ashy-white dolomite rocks, unrelieved by a single patch of green. In the valleys again the simple cherry and apple-tree have given way to the far more variegated and luxurious vegetation of a warmer zone, producing of course a greater diversity in colours than is created in the northern parts by the two or three shades of green peculiar to Alpine vegetation.

Gigantic chestnut and nut-trees, ivy-clad ruins and venerable old castles in a good state of pre-

servation in the foreground, with gardens and
vineyards, surmounted by ashy-toned cliffs, in the
background, are the characteristics of South Tyrolese scenery.

If, with regard to the Tyrolese themselves, the
experience of many years spent in Tyrol gives me
a right to express an opinion varying somewhat from
those of many authors, I must say that I have found
the Tyrolese in matters of daily life a highly intelligent, bold, and excessively hardworking people,
distinguished, even from the inhabitants of other
mountainous countries, by great patriotism and by
an innate unquenchable love for their native soil,
enhanced by a strangely chivalrous feeling of manly
independence. Regarding their warlike spirit—
fostered, to a great extent, by their strong attachment to the Hapsburg dynasty—we need but refer
to the endless wars in which the Tyrolese were involved from the very earliest times down to the
present day. In the Middle Ages the country was
hardly ever in a state of peace from external or
internal foes. Not only was it surrounded on four
sides by dire enemies, the Venetians, Italians, Swiss,
and Bavarians; but the broad Inn and the sunny
Adige valley, connected by one of the lowest passes
over the Alps, formed the chief high road between

civilised Italy and rough Germany. Not only was this highway, paved by Nature herself, used for commerce, accompanied, however, by a calamitous . system of rapacious highwaymanry, but it was also constantly crossed and recrossed by victorious or defeated armies marching to or returning from Italy. Whether these armies were hostile or friendly to the Tyrolese, the results were always disastrous to them.

There are, indeed, few countries that have suffered from war and its dire calamities so much as Tyrol, and though its affairs occupy but a small space in the history of Europe, yet to the student they afford quite as rich a field for his researches as the history of many a mighty and powerful kingdom.

Great heroism distinguished the Tyrolese on every occasion, generally indeed bringing them out the victors against odds. Their great power of endurance, superior muscular force, indomitable courage, and a certain love for fighting and hard knocks have, since the time when the generals of Charles V. and Maximilian recruited their best soldiers from the country, gained them high repute, quite apart from their deadly marksmanship, which even Napoleon's best generals and picked troops could not withstand.

Nothing demonstrates their innate love for their native soil more signally than the fact that, while in other countries a portion of the inhabitants emigrate to more propitious territories, a genuine Tyrolese very rarely indeed leaves his country for good. When their great purpose of life, the accumulation of small fortunes, as pedlars, musicians, or in other vocations, is accomplished, they never fail to return to their home, and, settling down in their native valley, enjoy the well-earned fruits of their industry.

There is something very pleasing in this attachment to the home soil, which carries a man steadfastly through difficulties, and incites him to overcome the ups and downs of a wandering life, and lands him at last, after twenty or five and twenty years' toil, in the promised land of his desires. It seems strange to meet in some remote corner of Tyrol men who, in the course of their constant travels, have acquired a certain polish of manners as well as a quite unlooked-for intelligence of thought and aptitude of expression.

To be addressed by one of these travelled Tyrolese, dressed may be in the very roughest of national costumes, perhaps even without a coat on his back or shoes to his feet, in the North German

dialect, or in French or English, is indeed sur-
prising.

Some of the men, particularly those who have
travelled in the character of Tyrolese singers, have
visited the four quarters of the globe. Many who
are known to me have exhibited their musical
talents at the courts of all the potentates of Europe,
and a few even in New York, Philadelphia, and San
Francisco. One of the latter, Ludwig Rainer,[1]
owner of a charming hotel on the beautiful shores
of the Achensee in Tyrol, related to me once his
various adventures while travelling in the United
States. He had been there three times. The first
time he fell into the hands of scoundrels who rid
him of every penny he had put by ; from the second
trip he returned not much the richer ; and only
the third time did he manage to amass the comfortable fortune he is reputed to possess.

Another man, now a well-to-do peasant, related
to me in capital English, interspersed however with
copious Yankee slang, how he had once been
blown up on a Mississippi steamboat ; while a third,
owner of a small inn in the Pusterthal, on my
asking him how he had come by his lacerated

[1] He and his troupe exhibited themselves, I think, on two
occasions before our Queen, and several times at the Paris and St.
Petersburg courts.

face, told me that while out bear-shooting in one of the Northern States of America, he had been suddenly attacked by a female bear, and not having time to draw his knife, he had succeeded in throttling the animal. The man's gigantic build and resolute demeanour was to me the best proof of his veracity.

The traveller who wanders through the Defferegger valley, a remote Alpine glen high up among the mountains, may, in certain months of the year, see a very singular sight.

The annual total emigration of the male population of this valley compels the women to do the work of the men. There is probably not a single man above eighteen or twenty, and below sixty or seventy years of age, in that valley for four of the spring and summer months.

You see women fell trees, drive their heavily-laden carts, till the ground, gather fodder, chop wood ; and if you enter one of the village inns you will see rows of women, their short pipes in their mouths, and elbows leaning on the table, drinking their pint of Tyrolese wine after their hard work.

A year or two ago I happened one Sunday evening to be present when one of the female occupants of the bar-room in the chief inn of St.

Jacob—I being the only man present—read to her companions a letter she had received that day from her husband, who at the time of writing was at Salt Lake City, among the Mormons. Though he was only a simple pedlar in hosiery, his graphic but inexpressibly quaint description of the city and of the customs of its inhabitants was highly amusing. Very singular and laughable it was to watch the effects of this description on the minds of the simple women, who had never heard of such a thing as the plurality of wives. Such a state of things seemed to them the height of human iniquity. Some thought the Mormons utter barbarians, while others, evidently applying the rule to their own homes, swore they would rather be killed than suffer any female rivals in their houses.

The Defferegger folk collect the necessary means to purchase their stock in trade by raising joint-stock companies. The man who contributes the largest sum of money to one of these modest commercial enterprises is also entitled to the proportinate amount of the net gains. They keep no books, nor have they any security in hand for the money invested; mutual confidence, engendered by a certain *esprit de corps*, with strict honesty among themselves, is the base upon which

these companies are built. In their business transactions with strangers while on their tours they
exhibit a sharpness quite unlooked for, and their
simple exterior and dull speech disguise in most
cases a very remarkable shrewdness.

Twenty or thirty years ago a very brisk and
remunerative cattle trade existed between two
Tyrolese valleys and Russia. The traders in this
business used to drive their droves of twenty or
thirty head themselves from Tyrol to Central and
Eastern Russia. When they could, they took advantage of a watercourse, as, for instance, down the
Danube to the Black Sea, thence along the coast
by land to Taganrog, and thence either north or
north-east. The large fairs at Nishnei Novgorod
and Orenburg were visited by them, and very
frequently they penetrated far into Asiatic Russia.
Their journey thither often occupied eight or nine
months, so that one venture entailed an absence
from home of eighteen months or two years. The
prices which they realised for the highly-prized
Tyrolese cattle used for breeding purposes were
naturally very high ; 500 ducats per head (about
250*l.*) was by no means an unusual figure for a
beast which they had bought in their native valley
for some eight or nine pounds.

The risks from accidents, disease, or natural causes were of course correspondingly high, and some men in one venture lost their all by the murrain destroying their drove, while others grew rich and prosperous in two or three expeditions of this kind.

Now all this is changed. The Russians are loth to pay fancy prices, and prefer getting their breeding cattle from England at a quarter of the former cost ; but it nevertheless gives us an idea of the intrepidity and commercial intelligence that prompted so highly venturesome and hazardous transactions.

Many a time have I been asked by some middle-aged rustic if I have ever been in Wolgsk, or Uralsk, or Orenburg, or Astrachan, and on my giving him a negative answer I have had to put up with the retort, 'then you have been no-where.' One or two villages in the two valleys that monopolised the Russian cattle-trade are entirely peopled by families who have grown rich in this trade, and who are now slowly descending the social ladder, step by step, till they reach the level of peasants, the stock from which they sprang seventy or eighty years ago.

The Tyrolese peasant has been often compared with a small freeholder in England, though of

course the latter, in comparison with a Tyrolese
cultivator, lives in the style of a prince or king.
A peasant proprietor who owns three or four acres
of tolerable land maintains himself and his family
in a simple but comfortable manner; he and his
son being sufficient for the labours of such a farm,
while his wife and daughters spin and make the
greater part of the family clothing.

There is however one very striking difference
in the circumstances of a small cultivator in
England and a peasant in Tyrol.

In the latter country all the cultivators are of
one and the same class, and therefore one has
the same chance as another; while in England
there are cultivators on a large scale able to apply
to the soil capital and skill with greater advantage
and economy than the small proprietor.

I have said that the Tyrolese exhibit a chivalrous
independence of character arising from an innate
confidence in their own powers. I might qualify
this observation by remarking that a kindly good-
natured courteousness towards the female sex, and
a bold, half defiant, half saucy bearing among
themselves are, generally speaking, marked characteristics of the young Tyrolese rustics.

The exuberance of animal spirits, the self-confi-

dence engendered by muscular strength, and the
jaunty, smart appearance of a young fellow dressed
out in his best, give him a sort of a 'cock of the
walk' air, increased by the fact that fighting is
looked upon by a young Tyrolese very much in
the same light as by a shillelah-swinging Irishman
on a visit to Donnybrook fair.

This defiant or saucy air generally sticks to a man
up to eight and twenty or thirty. Later on it is
supplanted by the natural results of an excessively
toilsome life, in the shape of a somewhat stern and
even morose expression of face. An angular, spare,
but well-knit and powerful frame replaces youthful agility and rounded forms. Hardworked as
women are in the Tyrol, their lot is by no means
an unenviable one. They are uniformly treated
in a kind manner by their husbands, and wife-
beating or brutal handling of women is entirely
unknown in the country. Their relation to man
in their spinster state reminds us in many points of
the chivalrous manners of society some five or six
hundred years ago. Morality is about on the same
par, and the lass who yields to the solicitations of
her lover who has proved his right in a fierce fight
with his rival or rivals, stands very much in the
position of the noble lady who, five centuries ago,

rewarded victory in combat and tournament with her love. The very poetry of the country is yet tinted with the sentiments of the 'Minnesänger.' What other people in Europe treat the whole subject of love in so quaint and charming a manner?

Nothing proves the vitality of this people more signally than the survival of the spirit of bygone days. Given to bouts of hard drinking, rough towards men, kindly in his manner to women, bold and warlike in his youth, cool and self-possessed in his age, the Tyrolese peasant, uncontaminated by civilisation, may be said to represent a strikingly true picture of a knight of the days of chivalry.

Poor and primitive as the Tyrolese are, and hardworking as they have to be, their lot is yet far preferable to that of many inhabitants of rural districts in Italy, France, England, and North Germany. The man, enjoying a life of domestic happiness, ignorant alike of real want and super-fluity, the woman, kindly treated by her husband, surrounded by healthy curly-headed children, can bear comparison with most, if not all, of the lower classes throughout Europe.

Of the defiant bearing that characterises the young folk, I may give one or two examples. A

custom very dear to a genuine Tyrolese is to adorn
his Sunday and fête-day hat with the tail-feathers
of the blackcock (*Tetrao tetrix*) and the 'Gamsbart,'
the long dark brown hair growing along that
animal's back at certain seasons of the year. The
tail-feathers of the blackcock are curved at the
extremity ; but if they are turned round so that the
curve or ' hook ' comes to be placed in a contrary
direction to that usually worn, a man is at once
metamorphosed from a peaceful native into a
quarrel-seeking ' Robbler.'

The manner in which a fight is brought about
by any young fellow stung by the Robbler's defiant
challenge is extremely simple. Stepping up to
him he asks, 'Was kost die Feder?' 'How much
for the feather?' the answer ' Fünf finger und
ein Griff' ('five fingers and a grip'), being
followed, before one has time to look round,
by a hasty rush and a fierce struggle, ending
frequently in bloodshed. Some fifteen or twenty
years ago, this practice prevailed throughout the
greater part of North Tyrol ; now, thanks to railways and tourists, it is confined to two or three
remote vales, where even at the present moment,
and I am speaking by experience, it is not safe for
a native of some other valley to sport a ' turned '

feather of the blackcock if he does not wish to invite a challenge.

I need hardly mention that the naturally quick eye of the Tyrolese detects at the first glance if a stranger, wearing a turned blackcock feather, is a Tyrolese or not. In the latter case the stranger can rest assured that were his hat garnished with twenty turned feathers no harm or insult of any kind would come to him. I have often been amused in watching the broad grin settling on the face, and mirth lighting up the eyes of a native as he sees a specimen of that most terrible species of continental tourists — some spindleshanked ' Berliner,' his ' pince-nez' on his nose, or a pale-faced, shrunken Saxon—strutting about with blackcock feathers on their hats, and displaying the invariable gamsbart —both, in nine cases out of ten, shams thrice over-paid—representing animals which these would-be sportsmen have never seen out of a zoological garden, much less shot.

The Zillerthal, which in my opinion, and in that of every traveller who has had occasion to see some of the really beautiful scenery to be found in other parts of Tyrol, scarcely deserves its fame for natural beauty, exhibited fifteen years ago—before it had been spoilt by the wide-spread repute of its landscape and quaint inhabitants—a curious medley of ancient and half-civilised customs. Among these institutions of the past was the 'Robbler,' or 'Haggler.'

The fact that a village could boast of a 'Robbler' of repute as its champion at fêtes or weddings was a matter of importance. If two such 'Robblers,' or even young fellows who claimed this honorary title, happened tó meet, or if one, hearing his rival's loud joddler, defiant and challenging to its last note, echo from mountain to mountain, he would hasten, guided by the sound of the repeated war-cries in the shape of joddlers, to the spot, where perhaps his foe was at work, and a fierce struggle for the supremacy in that part of the country would ensue. On these occasions severe injuries were the rule. A year or two ago an old wrestler, a famous robbler in his youth, died in his native village in the Zillerthal. The numerous disfiguring wounds on his body told the tale of many a fierce combat in his youth. His left eye, the better part of his nose, the tip of his ear, and two fingers were 'missing ;' he had also had an arm and a leg broken.

All this has now passed away. Such meetings, if they do occur, are decided by more legitimate means ; and certain laws and rules, strictly enforced by those present, confine the combat to the limits of a mere wrestling match. The use of the knife, at present even of frequent occurrence in the Highlands of Bavaria, was always discountenanced by the Tyrolese. Although the opinion may not be expressed in so many words, it is considered a cowardly act by the natives, and a man once caught in the act of lowering his hand while wrestling to the trouser-pocket from which the handle of the knife protrudes, is shunned henceforth, and any quarrel with him broken off.

Sunday or fête-day fights, originating in the Wirthshäuser, or village inns, now and then occur still. The usual cause of these fights is, of course, some buxom Helen, somewhat too free and indiscriminate in the display of her favours to her several admirers. It is obvious that the responsibilities of 'mine host' on Sunday and fête-day evenings, when wine and schnapps have done their work, are vastly increased.

A rural 'wirth' in Tyrol is a being it would require a whole book to depict with accuracy. A farmer himself, and owner perhaps of four or five horses, he is not only a man of importance in the village, but generally also of comparative wealth,

sure to be, or to have been once, at the head of the
'Vorstehung,' or municipality. He is '*the*' man who
dares to avow any anti-orthodox opinion in the
face of an enraged priest ; he heads the liberal party,
if there be any, in his village ; and his word very
frequently carries the day in any question of village
faction quarrel. Large, portly men generally, they
have to be firm and resolute, ' For,' as a giant ' wirth '
once remarked to me, 'a wirth who cannot expel
any one of his quarrelsome or drunken guests can
never hope to keep order in his house.' Though it
would be going too far to say that this is the rule,
the ' wirth's ' position is always one requiring men of
firm and determined character, who know, either by
their bodily strength or by their mental superiority,
how to make themselves respected and obeyed.

Nothing illustrates the stuff these men are
made of better than the important part they
played in the memorable war with the French.
Out of nine renowned leaders of the Tyrolese
peasant troops, no less than seven were ' wirthe : '
among them the Wallace of Tyrol, Andreas Hofer,
the ' Sandwirth,' as the populace term him.

Rare as fights are now, the customs which
rule these encounters nevertheless vary a good
deal according to the locality. In some valleys

the combatants content themselves with throwing each other ; in others, again, severe injuries are the rule. I once happened to be present in the Upper Zillerthal at a fight between four men. The ferocity of the combatants and the savage way in which they attacked each other rendered it amazing that no serious injuries were inflicted. An eye scooped out and two bleeding heads were about the only visible results. I was not a little struck with the cool and off-hand manner in which the victim of the first-named injury replaced his eye into the socket, to which it had remained attached by some fibres. A strip of cloth was bound over it, and the man rejoined his companions sitting round the table, all being the best friends in the world now that the quarrel was once settled. I may add that the loss of the eyesight is by no means the inevitable result of a 'scooped-out' eye, as long as it remains attached to the socket, and the nerves are not injured. I know a man whose right eye has been twice 'scooped,' and yet he sees perfectly well with it.

To give an idea of the hardships which fall to the lot of a Tyrolese peasant, I will endeavour to recount the odd features of some of the remote valleys noticed by me in the course of my wanderings.

In the Wild-Schönau (North Tyrol) not a few of the houses are built on such steep slopes that a heavy chain has to be laid round the houses and fastened to some firm object, a large tree, or boulder of rock, higher up. In many of the side valleys of the ' Pusterthal' manure and earth, the latter to replace the poor soil exhausted in one or two years, have to be carried up the precipitous slopes in large baskets, or ' kraksen,' on the backs of men. In one village off the Pusterthal, and in two others off the Oberinnthal, many of the villagers come to church with crampons [1] on their feet, the terribly steep slopes on which their huts are built, somewhat like a swallow's nest on a wall, requiring this precautionary measure, and they are so accustomed to wear them constantly on their feet during the week that on the Sunday they even come to church with them.

In Moos, a village not very far from the Brenner, having a population of 800 inhabitants, more than 300 men and women have been killed since 1758 by falls from the incredibly steep slopes upon which the pasturages of this village are

---

[1] A sort of iron sole, supplied with six or eight spikes, an inch or an inch and a half in length ; the irons are securely strapped to the shoe by means of leather or cord fastenings. They are of great help on precipitous slopes.

'WILDHEUER' RETURNING WITH PACK OF HAY.

situated. So steep are they, in fact, that only goats, and even they not everywhere, can be trusted to graze on them, and the hay for the larger cattle has to be cut and gathered by the hand of man.

The 'Wildheuer' is very numerously represented in the Tyrol. Their occupation is very similar to the one just described, with the difference that a 'wildheuer' climbs the highest eminences, up to eight and nine thousand feet, in search for the long Alpine grass growing on steep slopes. Armed with his crampons, he sets out on his dangerous task. If the precipices are too high to admit his precipitating the bundles of hay, closely packed in a sort of net, down the declivity, he has no other means of transporting it but to take the heavy burden, exceeding often a hundredweight, on his shoulder, and return by the same perilous path by which he ascended. So common in Tyrol are valleys having amazingly precipitous slopes, with not a patch of level ground in their whole stretch, that we frequently meet with proverbs quaintly illustrating the dangerous nature of a glen. Thus of one (Hochgallmig) the saying runs : ' Here the hens have to walk on crampons, and the cocks use Alpine poles.' Of another, ' If the swallows can't find any walls of suitable height in the rest of

Tyrol, they come to Taufers' (Oberinnthal) 'to build their nests on the slopes of the valley.'

In See, a tiny village in one of the remote glens off the latter valley the bodies of persons who had died in winter were formerly kept in the lofts of the houses till the snow vanished from the path traversing a mountain over 8,000 feet high, which connected See with the village to whose parish it belonged. See, however, with its population of 500 souls, has been recently added to a parish not requiring ten or twelve hours to be reached.

In another valley the letter-carrier, who visits it once a fortnight (in summer), is obliged to wear crampons on his feet for two days, and each day for more than twelve hours.

In many valleys the staple article of production is butter, which is carried over mountain paths to the next large village or town.

Thus in Hinter-Dux, about half of the male population of that valley are occupied during the summer months in transporting this commodity to Innsbruck. One of these men will carry 120 to 130 pounds, or about 150 English pounds, for eleven or twelve hours constantly on his back, and traverse two very steep ridges of mountains over which the path to Innsbruck, their market for butter, leads.

Considering the poor pay received by these carriers, and the exceptional fatigue attendant upon the transport of such a weight, it is astonishing that emigration is but rarely resorted to by natives of the Hinter-Dux and other valleys where similar precarious means of gaining a livelihood are the rule.

Strangers, oddly enough, very often find the unsophisticated population of the remoter parts of the country the most difficult to deal with. This is caused to a great extent by the suspicious shyness ·with which these rustics glance at the strangely-dressed invader. Nothing aids one's efforts· to penetrate the outer coat of reserve, and at the same time to gain a true insight into the lives and characters of this people, as an assimilation to their habits, customs, language, and dress. But very naturally too, as all travellers do not care to acquire the necessary broad German, or to walk about in short ' leathers ' with an old hat on one's head, I must content myself with asking the reader to make his own inferences from the following sketches of Tyrolese life.

I may as well mention here that my adoption of the native dress and language has very frequently been the source of great amusement to me.

A worn shooting-jacket on the back, with short, time-stained 'leathers,' displaying a bronzed knee, is an apparel that not only opens the hearts of the natives, but also the minds of unsuspicious tourists.

Many of my readers no doubt will know the exquisite view from the 'Matreier Thörl'—a pass intervening between the two villages of Matrei and Kals in the Tyrol. On a fine August day, two or three years ago, I was lying at full length on the short grass, basking in the warm afternoon sun, on the height of this pass. A three days' unsuccessful chamois-stalking expedition high up among the opposite range of snowy peaks had brought me on my return to civilised quarters across this height. Feeling rather tired, I determined to wile away a few hours till approaching dusk would render a speedy descent to Kals advisable—for that day my goal. I had not been more than half-an-hour thus enjoying the grand view and the absolute and impressive tranquillity reigning around me, when I perceived a group of tourists slowly climbing the narrow path leading to the celebrated point of view, on the height of the 'Joch,' or pass.

Retreating to a patch of rhododendrons a few yards off, in order to be out of the way of the puffing and 'winded' tourists, I immediately learnt on their arrival, by the 'charmings,' and 'delightful,' and 'beautiful,' that fell from the lips of the three ladies that made up the female contingent of the group, that the guess which I had made on first seeing the group, when yet half a mile distant, was right.

An hour or so was spent by the party in admiring the view, sketching the valley at their feet, and deriving animal comfort from sundry parcels and bottles produced from the knapsacks of the two men, one evidently the father, the other the son, and apparently a university man. The fact that they were unprovided with guides or porters was explained in the course of their conversation by the casual remark of one of the ladies that they hoped their luggage had safely reached Kals, the village they were intending to gain that evening.

Not wishing to play the eavesdropper any longer, I had swung my 'Rücksack' on to my shoulders, and was just taking up my rifle in order to turn my steps Kals-ward, when a hasty exclamation of one of the younger ladies, to the purpose that she desired to sketch me as representing a typical Tyrolese chamois-hunter, made me hasten away. The brother, evidently the only one of the party acquainted with German, ran after me, intending to secure me as a model for his sister. The excuse—in German, of course—that I was pressed for time, and had a walk of two or three hours' before me, got rid of this proposal, only, however, to get me into a worse scrape. Asking me if I was going to Kals, he seemed quite astonished to hear that it was nearly three hours off, whereupon he informed his relatives of the unwelcome piece of information gleaned from 'this fellow,' pointing to me. Hardly able to suppress my laughter, but desiring to retain my incognito, I was just going to pass on, when my interrogator asked me in his execrable German if I would mind showing them the way down. My hint that the path could scarcely be missed was met by the further request of the ladies that I would carry their shawls, which had thus far been fastened to their waists by straps. Escape seemed impossible, and not wishing to be disobliging or uncivil, I assented. Ten minutes later I was stalking in front of the file, now rid of their shawls and knapsacks. The latter had been introduced into my spacious 'Rücksack' by the young man, who imagined that I had not observed the addition of weight. 'These fellows don't feel fifteen or twenty pounds more or less on their backs,' was the off-hand speech with which he quieted the remonstrance of one of his sisters.

Close behind me tripped the two girls, the parents in the centre, and the son closing the file. The confidential conversation of the two young ladies, both bright and handsome specimens of that most pleasing of England's characteristics—her fair sex·—to which I had to listen for two long hours, must of course remain untold in these pages ; let it suffice that the concoction of a strategical device how to get me into their sketch-books, intermingled with personal remarks, not uniformly flattering, on my humble self's appearance, formed the chief subject of their constant chatter, making me rejoice that the even path and their sure-footedness rendered the extension of a helping hand to the two fair conspirators unnecessary. Just before dark we reached the straggling village of Kals, and the 'Gasthaus,' a modest, but scrupulously clean little inn.

Dreading to enter the house in the character of a porter, as I was well known to the host and the guides, who were sure to be lingering about the entrance, I came to a sudden halt a few yards from the inn. Unfastening the knapsacks and bundle of shawls from my 'Rücksack,' with the intention of handing them to the two gentlemen of the party,

I meant to make off to another little inn, where I hoped to be safe from any unwelcome *dénouement.*

An ominous whispering, and the accompanying jingle of loose money, made me recollect that my 'porter' character entitled me to a fee. 'Here, my good fellow, are two florins for your pains,' were the last words I heard, for with a sudden turn I was off, leaving the 'paterfamilias' rooted to the ground with outstretched hand. Fate, however, meant differently, for with a slap on my shoulder, and 'Why, my dear Mr. Grohman, where on earth are you off to in such a hurry?' I was brought to a dead stop, not five yards from my bewildered 'employers.'

A London barrister, whom I had accidentally met some weeks before while on a mountaineering tour in the Dolomites, was thus destined to tear off my porter disguise, and, what was far more disagreeable, made me the object of profound excuses on the part of my late 'masters.' Of the blushes of the two charming conspirators on seeing the Tyrolese chamois-hunter transformed into a fellow-countryman, whom they had unwittingly made their confidant on more than one point, it is unnecessary to speak ; nor of the upshot of the whole mystification, a charming supper in the little parlour of the inn, and a far more charming tour in their company back to Lienz, and into the heart of the Dolomites, followed, five or six months later, by several very merry dinners in a certain house not a hundred miles from Hyde Park Corner.

On another occasion—for this incident recalls to my mind a host of ludicrous scenes,—while sitting at a crowded dinner table in Schluderbach, near Ampezzo, and chatting with a stout old monk, I had to lend an unwilling ear to some very severe criticisms on the part of two somewhat emancipated English ladies of a certain age, on the beastly custom of my stout neighbour of indulging in very frequent doses of snuff; and then, when that subject was exhausted, to no less stinging remarks on my own appearance. A flannel shirt and a shooting-jacket of Tyrolese cut are perhaps not the guise in which I should care to appear at a Swiss *table d'hôte*; but for the primitive Tyrolese hostelries, those two ladies exercised, I am inclined to think, somewhat too harsh a judgment.

For the benefit of those of my readers who have never had occasion to cross the threshold of an Alp-hut or châlet, I may add the following short sketch of these elevated summer abodes of vast numbers of Tyrolese. In May, when the last streaks of snow have vanished from the mountains of medium height, the peasants, now rid of their autumnal stock of fodder, lead their herds of cattle up to the juicy pasturages on the mountain slopes that encircle their native valleys. These 'Alps' or pasturages are resorted to at different seasons, according to their heights, and many of them, at an elevation of 6,000 and 7,000 feet above the level of the sea, afford the necessary food for the cattle only for a short period.

Each pasturage is provided with a hut, the châlet or Alp-hut, and a rich peasant will tell you that he has three and four of these 'Alps,' situated one above the other at an interval of an hour or more between each. Thus when the grass on the lowest, which is first resorted to, grows scarce, the herd and his cattle migrate to the one higher up, and in this way the highest Alp-hut is reached in the warmest season of the year, about the month of July.

Poorer peasants have two Alps, and if the peasant has but a few head of cattle to call his own he will be even content with one, though this may be said to be the exception in all but the very poorest valleys.

The Alp-huts are simple log-huts divided into

two unequal divisions. The larger part at the rear provides the necessary shelter for young cattle in wet or cold weather, while the smaller front portion is the kitchen, parlour, and bedroom of the man or woman to whose guardianship the cattle are entrusted. On mountains abounding with grassy slopes we find clusters of these huts together, often to the number of twenty or thirty.

The interior of these huts is extremely primitive. The fireplace occupies one of the corners, and is generally a sort of pit or trench, dug around by way of a seat, surmounted by a crane, from which is suspended the huge black caldron or kettle, the most necessary utensil for the manufacture of cheese.

In large and prosperous Alp-huts these caldrons are of amazing size, and I well remember that in my younger days it was my habit at night while sojourning in these châlets, to seek a warm though somewhat confined resting-place in the inside of one of these giant kettles. Once, in fact, I was nigh drowned, by the ' Senner,' or cowherd, pouring a huge pailful of water into the caldron, ignorant as he was of its contents.

In Styria, Upper Austria, Salzburg, and certain valleys in Tyrol, girls,—strong, healthy-looking lasses—are the occupants of these solitary huts, while in other parts of Tyrol and in Switzerland a man guards the cattle entrusted to him. If the peasant to whom the Alp belongs is unable to afford to keep such a ' Senner ' or ' Sennerin,' his grown-up son or daughter, as the case may be, is sent up in that character.

These people have but little opportunity of indulging in that Arcadian leisure which romance assigns to tenants of solitary Alp-huts. The manufacture of cheese, the churning of butter, the milking of the cows twice a day, the cleaning and arrangement of the dairy utensils, and the responsibility of keeping their flock from straying into dangerous places, and attending on sick cattle, give them constant and excessively arduous occupation.

A bed of straw and a blanket on a sort of projecting balcony in the inside of the hut is their resting-place, and the stranger or native who seeks a night's shelter has to content himself with the fragrant hay on the loft right over the second partition, where the cattle seek a welcome shelter from the inclemencies of a rough Alpine climate.

The dairy or milk-cellar is either underground or in a small chamber off the front division. As

A SATURDAY EVENING CONCERT ON HIGH.

the type of châlet in which the Senner is the pre-
siding master has been often described in books on
Swiss travel, I shall confine myself to the more
preferable class governed by female hands.

Greater cleanliness in dairy matters, the generally scrupulously clean interior of the hut itself, and
the far more pleasing and attractive welcome ac-
corded to the stranger, are some of the manifold
merits of the latter custom. Little more than a
hundred years ago the Senner was an unknown
being ; every Alp-hut in the Tyrol was presided over
by Sennerinnen. The Archbishop of Salzburg, to
whose diocese many of the Tyrolese valleys ap-
pertained, moved by sundry complaints respecting
the somewhat profligate life led 'on high,' gave
strict injunctions that henceforth no 'Sennerin'
should be allowed. The Bishops of Trent and
Brixen followed suit, though not in so rigorous a
manner. Since that time, however, and chiefly
since the wars in the first years of this century, the
buxom, healthy-looking Alp-girl has reoccupied her
former position in not a few Tyrolese valleys.

Saturday evening is the grand 'reception' night
of these gay and merry lasses. Work over in the
distant valley, each young fellow who is lucky
enough to be able to sing : 'A rifle on my back, a

buck chamois in my bag, and a black-eyed, merry
Alp-girl in my heart,' takes his rifle, his scant stock
of provisions, and is off to the Alp-hut high up on
the mountains, where he knows his lass is awaiting
him. Far off, while the low châlet is yet but a
speck, a piercing, echoing 'joddler' of the lover
will bring his lass to the door, and a minute later
a sharp silvery answer will float down to the
mountaineer, whose feet cover the intervening distance with a speed that love only can accomplish.

Sunday is devoted to stalking or poaching, and
on Monday morning, long before daybreak often,
the swain is off in order to regain the site of his
daily labour by five o'clock, the hour for beginning
work.

Playing the Don Juan is not unfrequently dangerous work for a stranger or a native of another
valley, and I have come across several instances
where a speedy retribution overtook the pirate in
strange waters.

In October and in cold autumns, when snow
falls in September, often even sooner, the Alp-girl,
aided by a peasant or a boy, returns with her
twenty or thirty head of cattle to the home valley.
Tinkling bells, hung round each cow's neck by
broad leather belts, wreaths of flowers, loud rejoic-

ings mark this event; and lucky is the fair lass who has made her allotted quantity of cheese, churned the requisite hundredweights of butter, and brought back her flock without accident or mishap to any of them.

In a closing remark to this introductory chapter, I wish to draw the reader's attention to another peculiarity of the Tyrolese. It is the creative genius that has distinguished this people for centuries. Painters, carvers, poets, musicians of repute form the body of the Tyrolese contingent of celebrated or well-known names.

Musical talent is, without comparison, the gift of nature most widely diffused in Tyrol; and to a stranger, particularly an Englishman, it is amazing to find a finely developed ear and a capital voice in the commonest country lout, who scarcely knows his A B C, and to whom Bismarck is an unknown being. To be able to join with a second or third voice in a song which they have not heard before is a very common accomplishment. Often have I been amused by watching the expressive face of some country lass listening for the first time in her life to the full tones of a piano.

To give an instance of this fine sense of music: a lady of my acquaintance was one afternoon playing and singing a Viennese air. The windows of the room were open, and two country lasses passing along the road stopped and listened for a little time. Presently, when at my request my friend repeated the song, the two girls fell in, one with the second and the other with the third voice. Being a stranger to Tyrol, my friend would not believe that the girls were common peasant lasses, unacquainted with the piece of music which she played ; and so, in order to convince her, I sent down for them and made them accompany her in a number of songs which she sang to try them. Their intonation and expressive voices excited her admiration, no less than did the piano that of the buxom lasses. My reader must not imagine, however, that the Tyrolese are fond of exhibiting their innate talent for music. Stubbornly shy, they will often refuse to sing any of their national lays if they see that their listeners are strangers. Tourists who keep to the frequented highroads, following the ruck of travellers, will hardly ever hear a genuine Tyrolese song. To enjoy a musical treat of this kind we must leave the carriage-roads and strike into the more unfrequented paths, and if possible visit remote Alp-huts. If we do not press the ' Senner ' or ' Sennerin,' or betray by any sign our wish to hear them sing, it is probable they will begin of their own accord.

Sitting on the low step in front of her châlet, enjoying a quiet half-an-hour's rest in the calm evening after her fatiguing day's work, the ' Sennerin' will awake the echoes of the surrounding heights, answered, perhaps, if there be other huts within earshot, by their inmates. Tinkling bells, the rich silvery voice melodiously tender in all its notes, the quiet calm of the evening, and the grand landscape, all unite in producing an effect that will remain impressed upon the mind for many a day to come.

I may here remark that the Tyrolese entertain a passionate love for the mimic art. The famous ' Mystery Plays' of the Middle Ages are supplanted by the modern 'Passion Plays,' organised on the same principles as those at Ober-Ammergau, though in most cases on a much smaller scale. Theatrical representations of all descriptions are highly patronised. Of the many I have had occasion to visit, I remember in particular one— given in a small village near Kufstein—bearing the title ' Richard, King of England, or the Lovers'

Tomb.' My mirth was great when, as an appro-
priate finish up of the cruel king—the chief character—his head is bitten off by a make-belief lion,
while a chorus, consisting of three peasant boys
and two lasses, yelled out, 'Thus perish all cruel
monarchs!'

CHAPTER II.

PRIESTHOOD AND SUPERSTITION.

To the fact that Tyrol is the most exclusively mountainous country in Europe—even Switzerland containing a larger relative proportion of open country—we must attribute most of the peculiarities and customs that strike the observer, and to some of which we have referred in the previous chapter.

To one of the most important characteristics I have, however, not yet drawn my reader's attention. It is the exceptional position of the clergy. Tyrol, one of the strongholds of the Roman Catholic faith, is ruled to an astonishing extent by the priesthood; and though in the course of the last ten or fifteen years the Church has lost a good deal of her former influence and power in the three or four larger valleys of North Tyrol, the ignorant natives of the more secluded and poorer Alpine

glens are yet terribly in the clutches of the
'Blacks'—the name given to bigoted priests.
Superstition and blind belief in the power of
their Church are the two firm rocks upon which
the clergy have erected their structure of spiritual
government, leaving the civil form of judicature
far behind in importance and energetic vigilance.
In a country where social laws are yet at a low
degree of development, reminding us only too
often of customs and habits of the Middle Ages, we
must be glad that any power exists able to curb
the animal passions of a primitive people. At the
present moment (and I have no doubt he will do so
for many years to come), a peasant dreads the
punishment inflicted by his priest—consisting of
perhaps a temporary refusal to grant absolution—a
hundred times more than any fine or sentence of
imprisonment which the law can inflict upon him.
What is a month's imprisonment to a man whose
mind is overcharged with the horrible pictures of
hell and the everlasting tortures which are sure
to follow disobedience to the ordinances and laws
of the holy Catholic Church?

I have hinted at the low scale of morality of the
Tyrolese, and without entering into any unpleasant
details, it must be remarked that among the lower

classes of the population the intercourse between the sexes is decidedly freer than in most other countries of Europe.

There are two or three conspicuous causes to which we can trace this. The most prominent are the municipal restrictions that cumber marriage among the lower classes in the rural districts. Very recently, only, has the Austrian Government annulled the law which compelled a man, desirous of entering into the holy bonds of marriage, to prove a certain income, and further, be the owner of a house or homestead of some kind, before the licence was granted. The heads of the parishes, very naturally too, gave the necessary permission reluctantly if they entertained the slightest fear of having ultimately a pauper family thrown upon the poor resources of the parish. Owing to this, and to the fact that nearly 40,000 Tyrolese, generally young men, leave their country every year in search of employment which keeps them away from their homes for the better part of the year, the majority of couples contracting marriage in Tyrol have passed the meridian of youth.

Next in importance, as a cause, is the lax way in which the Church deals with licentious miscon-duct. Strict in most vital points, she shows a re-

markable deficiency of energy in combating with
an evil, which, it is true, does not touch the in-
terests of the Church herself, but yet would be
worthy of her most strenuous efforts to abolish.
Immoral intercourse between the sexes is, in her
eyes, a minor iniquity, expiated by confession. We
must remember, too, that the conduct of the priests
themselves is not infrequently open to the severest
criticism. Free as the intercourse between the
sexes is, we have nevertheless to note one redeem-
ing quality, the sacred light in which the marriage
vows are held. Unrestrained as a woman's career
may have been before her marriage, she becomes
a dutiful, hardworking wife when once the holy
knot is tied.

As in certain rural districts of England (the
North and West), where formerly women usually
refrained from marrying until they were on the eve
of becoming mothers, we find that on an average
half of the wives of Tyrolese peasants have had
children before their wedding-day ; and though it
is quite true that the lover very rarely forsakes
the mother of his illegitimate offspring, and ultimately marries her, we must not ascribe this final
act of justice solely to the good feelings of the
male culprit, but rather to the power of the priest

over the mind of the sinner confessing his guilt. The priest it is who urges him to set right an old wrong by marrying the girl, who but for the absence of the holy bond was to all purposes his wife ; and were it not for his lively pictures of ever-lasting tortures in a certain subterranean abode of sinners, the percentage of girls abandoned by their lovers would be far greater than it is.

As in most Roman Catholic countries, the Church in Tyrol counts her most effective and devout disciples and followers among the female portion of the inhabitants. The simple and credu-lous mind of the ignorant peasant woman acts as one of the mainstays and supports of the whole structure of absolution, redemption, or, on the con-trary, eternal damnation, one and all dependent upon the volition of a mortal man, her priest.

It is only in the course of the last twenty or thirty years that the custom, spread throughout the country, of 'Fensterln' or 'Gasselgehen'— the introduction of the lover into the bedroom of his lass—has been stopped in the three or four larger valleys, while in the rest it flourishes to this day.

Priests have told me that thirty years ago the custom of sleeping in an entirely nude state, and

crowding all the members of the family into one
bedroom, was the constant theme of their dis-
courses from the pulpit; and even now-a-days I
have frequently listened to sermons of some well-
meaning rural priest, the subject of which was the
necessity of washing every day and changing one's
linen once a week. Well aware that sentiments
of propriety are foreign to the minds of his
listeners, the priest does not base his exhortations
on the supposition that a clean face once a day
nd a clean shirt once a week are domestic comforts
necessary to the equanimity of the human mind, but
rather on the consideration that a dirty face and filthy
shirt are obstacles in the path of true love. ' For
how,' I once heard a loud-voiced rural priest hold
forth, ' can a comely girl feel herself honoured with
the love of a man approaching her in dirt-begrimed
clothes, emitting an effluvium sufficient to knock
a man down at ten paces ? ' The worthy pastor was
in this instance urging the necessity of abolishing
that filthy custom of the male cowherds, who in
the beginning of the summer leave their native
village for the more elevated pasturages, and return
with their cattle in autumn, having the same shirt,
unwashed the whole five or six months, on their
backs. The dirtier and thicker the coat of filth on

the shirt, the more honourable for the wearer, for does it not speak for itself, that the owner has been in the meantime busy and hardworked? This custom, I am happy to say, is confined to those valleys where male cowherds are sent up to the Alpine pasturages, and it is now fast disappearing. It is in this way that the priest attains his object, and hundreds of instances could I recite of this indirect and roundabout manner of overcoming prejudices deeply rooted in the hearts of the people.

Thirty or forty years ago brutal and sanguinary fights between rivals in the love of one and the same girl were the invariable finish-up of fêtes, weddings, christenings, and, in fact, all assemblies. The loss of the nose, an ear, or a couple of fingers, bitten off by his foe, marked the vanquished for life. The still more brutal act of scooping out a foe's eye—by a jerk of the thumb—was at one time a very prevalent abuse, and even nowadays in one or two valleys this barbarous habit still exists, though, thanks to the strenuous efforts of the clergy, it is far less often practised. Among the several more or less mischievous results entailed by the great supremacy of the clergy, the gross

superstition and devout belief in their supernatural
powers are about the most harmful.

The two following instances are sufficient to
substantiate my statement and show how solicit-
ously a Tyrolese priest will 'dress up' some
commonplace event in the garb of a semi-
miracle, and how by hook or by crook he manages
to impress his parishioners with his power to charm
evil spirits,

Two years ago a certain deformed tailor in the
village of Vomp (near Schwaz, in the ' Unter-Innthal ') was attacked by a somewhat violent fit of
delirium tremens, brought on by too liberal pota-
tions of spirits the day before.   His family, terribly
frightened by this hitherto unknown malady, sent
for the village doctor.   After a protracted examina-
tion of the patient this most enlightened disciple of
Æsculapius declared himself incompetent to deal
with the mysterious ailment.   All he could do was
to advise the immediate attendance of the priest.

This piece of advice was of course promptly
followed, and ten minutes later the priest in his
official capacity, attended by two acolytes with
swinging censer and holy water vessel and mop,
was standing at the bedside of the raving hunch-
back.

Grand opportunity to work a miracle, thought the holy man, and forthwith the solemn declaration that the patient was possessed of the devil made the assembled household and the mob standing outside the house shake and tremble in their shoes. The room was cleared of the gaping and frightened crowd, and the priest began his course of recondite exorcising manipulations, an interesting description of which is furnished in the following literal translation of an account (which appeared in one of the most popular local newspapers) of the further proceedings of the devil while closeted in the confines of a narrow chamber with a priest, armed with rosary and censer. I have unfortunately to refer my readers to this piece of second-hand information, as very naturally no mortal but a clever editor could have penetrated the veil of mystery that clung round that dire eight hours' struggle.

'After four hours of uninterrupted praying and declamation of Latin adjurations and exhortations that filled a handy "Benedictiones" prepared for like occasions, the holy man, faint with hunger, proposed to leave the devil for an hour or so in undisputed possession of the tailor, while he, the holy, but mortal man, ate his dinner. This intention,

E

however, was not carried out, for with a hellish
peal of scornful laughter the evil spirit informed
him that if he left, he—the Satanic Majesty—
would take perpetual possession of his victim.
This threat of course needed a firm answer, and so
with renewed vigour the holy man continued his
exorcising.

'Four hours more of Latin formularies, hailed
down hard and fast upon the devil-possessed
patient, at last brought his Hellish Majesty to bay,
and with one discordant whoop of defiance the evil
visitor took his departure through the window
opened by the priest for this purpose.

'The priest, eager to close the casement, and
thus to make a return of his vile tormentor impossible, reached the window, and was just about to
shut it when a large dog, lying in the courtyard of
the house, set up a howl, thereby indicating very
plainly that the Devil, unsuccessful in other
quarters, was determined to get somebody or
something to accompany him to his hellish retreat.
A rifle in the hands of the master of the house
speedily put an end to the dog's existence, and thus
his Satantic Majesty was deprived even of his
canine victim.

Eight hours of unremitting exhortation were

needed to drive the Evil Spirit from that God-forsaken house.

'As soon as the miraculous success of this priest became known to the crowd surrounding the house, loud rejoicings and fervent prayers were offered up.'

The next Sunday this event was grandly dilated upon from the pulpit, and after service numbers of holy pictures, representing the heart of Jesus, wreathed round by suitable verses and hymns, were distributed among the parishioners.

These holy amulets against a second visit of the devil were nailed to the house-door, stable-door, and barn-door of every house in that village, and since then the population have enjoyed a blissful security from his Satanic Majesty. For the truth of this event in all its details, save those of course that occurred in the sick room, I can vouch, as I was present and saw most of the proceedings myself. The exact date June 23, 1873. Not so bad for the nineteenth century, my readers will exclaim.

The second instance is much simpler and far less wonderful.

A peasant whose fields were infested with the grub of the cockchafer (they remain three years in

E 2

their caterpillar state, appearing in the fourth as chafers) complained to the priest of his village of the nuisance, and asked his advice how to get rid of them. It seems that they had already been doing grievous damage to his wheat and corn for three years, and the priest on hearing these details found himself induced to promise their expulsion from his parishioner's fields. The promise of a couple of sacks of corn and a huge wax candle to the Holy Virgin no doubt had something to do with the priest's readiness to comply with the peasant's request. Two acolytes, a basin of holy water, a huge mop wherewith to sprinkle the fields, and some incense, were all that was needed. On the termination of the priest's promenade round the ground (his holy book in his hand and two acolytes swinging the censers in front of him) he declared that next spring the grubs would fly away.

And really, wonderful to say, next year the creeping grubs took wing (as cockchafers), leaving the happy owner of their playground during the last three summers to his meditations on the miraculous power of holy water and incense in the hands of his priest.

A recent able authoress [1] has given a rich store

[1] 'The Valleys of Tirol,' by Miss R. H. Busk.

of myths, superstitions, and interesting instances
of what the Germans call 'Volksaberglaube,' the
superstition of the populace in Tyrol; but there
still remain in the remote parts of the country odd
customs displaying a devout belief in good and
evil spirits, national traits which, with one or two
exceptions, have not yet found their way into
English, nor, so far as I am aware, into German
works upon Tyrol. Looking down the long list of
these customs—we might call them relics of the
past—I find that most of them represent precautionary measures against evil spirits in general and
the devil in particular. I must premise that a
Tyrolese peasant never mentions the word 'Teufel;'
to him any word is better than 'Devil.' We therefore find him called the Evil One, the Black One,
the Bad Spirit, or the 'Damned One;' and even
the low oaths used by the Tyrolese are conspicuous
by the absence of the word, which in English,
French, German, and most other languages is a
common imprecation. I do not by any means put
this forward as a laudable characteristic of the
Tyrolese, for, like other Roman Catholics, they
will make profane use of a Name which, according
to our English feelings, is not to be called in vain.

I merely mean to say, that just as the common

Tyrolese does not make the slightest difference
between Protestant and Jew, but terms every non-
Roman Catholic a Jew, the shunning of the word
' devil' illustrates in a remarkable manner that
dense ignorance on religious matters, which is
deemed by the clergy the best safeguard against
any repetition of those dangerous revolutions in
religious matters which on one or two occasions
were near overthrowing the old faith. Not once,
but a hundred times, have I been struck by the
uneasy glance around and behind him, when, in
joke, I have mentioned the word ' devil ' to a rustic
inhabitant of some remote little village. The sign
of the cross and a hasty ejaculatory prayer are on
such occasions supposed to be the only preserva-
tives against an immediate appearance of the Evil
One himself !

The Tyrolese peasant connects every elementary visitation, such as hailstorms, lightning, earth-
quakes, heavy rains, or long droughts, with the evil
disposition of the Unholy One, or sees in it the
punishment for some unrighteous act.

Before he sows his field he sprinkles it with
small bits of charcoal consecrated by the priest.
When he drives his cattle to the mountains, his
Alp-hut receives the blessing of the holy man.
When his cow calves she is besprinkled with

holy water ; before he enters an untenanted house he goes over his rosary. When a thunderstorm is approaching the village bells are rung, and if he has a bell on his house—well-to-do peasants in the fertile valleys very often hang a bell on top of their house, to call to their meals their men and women servants from their work in the fields—it is set tolling with might and main. The object of the ringing is to keep off or charm the dreaded lightning. The peasant population have in this safeguard a staunch belief, which is not shaken even if the lightning strikes that or any adjacent house. ' The bell has been bewitched,' they argue, 'and requires to be re-consecrated.'

As a rule the older the bell of chapel or church the more efficacious it is considered, and one or two in different parts of the country have a wide-spread repute as ' Wetterglocke,' or storm-bells. You often will hear a peasant express regret that his village possesses a bell much inferior to that of the next village, and adds, ' Oh, had we only the bell of Rodenegg!'—a bell enjoying the highest repute as a lightning charmer throughout Tyrol.

To touch a person killed by lightning before the priest has spoken a short prayer over the body is considered highly dangerous.

To counteract the devastating results of a heavy

hailstorm, a bunch of twigs of the round-leafed willow, duly consecrated on Palm Sunday by the village priest, is stuck on a pole in the middle of the field.

On Christmas Eve every door in a peasant's house is marked with three small crosses in chalk, 'to keep out the Evil One,' as they would tell you if you asked why.

When a woodcutter fells a tree slightly injured by lightning he immediately cuts three crosses on the level surface of the stump.

To wash a child before its forehead has been touched ·by holy water (two or three small vessels filled with it are never lacking in a peasant's dwelling) is highly injurious to it.

To pass a chapel, roadside shrine, or cross, or the wooden beam adorned with a votive tablet, without making the sign of the cross, or taking off your hat, is considered by the peasants as highly improper, and I have known men turn round upon me with an expression of anger or astonishment depicted upon their faces when they remarked my non-observance of this custom.

To give an instance of the peasant's superstition respecting lightning, I may relate here an incident that occurred to me a year or two ago.

In a small and remote village, consisting of nine or ten houses and a small chapel, the priest of the next village, some hours off, used to read an occasional mass for the benefit of the weak and decrepit who were unable to attend the distant place of worship. In this chapel I had discovered four very remarkable pictures of sacred subjects painted evidently by an old German master of repute.

Though eager to purchase them, I knew my customer too well to show any great wish to possess them, but broached the subject by offering four new pictures in their stead. My offer was refused, and it was only after I had doubled the price I had previously offered, and promised to pay for the restoration, viz. whitewashing, of the chapel, that the owner of the edifice would hear of parting with the dusty, hardly visible old paintings.

A week later I had returned to the village accompanied by four men, who carried the pictures which I had bought in the meantime in Innsbruck.

Hardly had I entered the peasant's house when to my utter astonishment he told me that he could not possibly part with the paintings I desired so much to possess.

After a considerable time spent in talking, I discovered at last the cause of the sudden refusal.

It seems that for many years lightning had never struck individual or house in that village—it occupied a very elevated plateau, and was therefore somewhat exposed to lightning—and now that his neighbours had heard of this proposed exchange they had united their voices to urge him not to part with them. 'It is just these pictures which may have preserved house and human being hitherto from lightning,' my uncomfortably superstitious vendor informed me. All talk on the matter was useless, so as a last remedy I assembled the whole nine or ten peasants that evening in the wainscotted low-roofed chief room of the owner of the chapel. My persuasive powers however again proved useless, and next day I had to return to more civilised quarters, carrying the new pictures back with me. Naturally I was greatly vexed at my disappointment and the loss of the money spent on the pictures, which now—they all represented gaudily-painted saints, or the Virgin Mary in various poses in heavy gilt frames—were for the time quite useless. Fortunately, however, I kept them, and did not give them away, as I had intended, for hardly six months later a flash of lightning fired a house in the village and killed several head of cattle. On hearing of this mishap I knew I had won the game, and a few days later I was in possession of my prizes.

Had I got the pictures the first time the peasants would have said, of course, that my exchange had brought about this untoward event.

In Ultenthal—to give an instance or two of the belief in local legends—there exist at the present moment the ruins of the strong feudal castle of Braunsberg, founded by a noble of that name in the early part of the twelfth century. A descendant of the founder, Knight Henry, took a part in one of the crusades of that century, and while on his perilous expedition, undertaken, as we may suppose, for the redemption of a soul laden with a long list of dark crimes, he entrusted his beautiful wife Jutta to the care and protection of his steward.

The latter, handsome Gunibert, proved himself a shameless Don Juan. The virtue, however, of fair Jutta, somewhat exceptional in those days, was deeply ingrafted upon her nature, and his subtle schemes only made him the object of her scorn and disgust.

Learning that his master, Knight Henry, had
returned from his dangerous voyage, and was but
a day's journey from his castle, Gunibert entered
his mistress's chamber and ruthlessly tore from her
fair hand the gage of love, the wedding-ring.

Mounting a fleet steed, he left the castle and
met the returning hero at the beginning of the
valley. Producing the ring, he told him a tale of
such base and calumnious defamation of his wife's
virtue that the enraged Count swore he would cut
off her head.

Jutta, troubled in her mind, and uncertain what
to make of Gunibert's violence, mounted the steps
of the high watch-tower, overhanging a terrible
abyss, at the bottom of which a turbulent torrent
boiled and seethed.

All of a sudden she perceived a large train of
armour-clad nobles and men-at-arms, headed by her
husband, riding up the steep incline leading to the
gate. At the side of the latter rode brazen-faced
Gunibert, evidently bent upon impressing his
noble master with the truth of certain facts.

Her quick eye guessed the whole truth of the
faithless retainer's revenge, and with a piercing cry
she precipitated herself from the giddy height into
the dark abyss at the foot of the tower. Wonderful

to say, she remained hanging on a bush which none had ever noticed before, overlapping the caldron of foaming water. The Count and Gunibert, riding up to the brink of the precipice, saw her thus suspended, and the latter, stricken by the hand of God, threw himself into the water hundreds of feet below him.

Even now, more than six hundred years after this tragic event, a blue flame marks the spot where the treacherous villain was drowned. Beautiful and faithful Jutta, saved in so wonderful manner by the hand of God, accompanied by her pious husband, who was overcome by the benevolence of his Creator, left the castle and entered the cloister of Weingarten, in Bavaria, where they ended their days in a manner befitting this remarkable event in their lives.

The origin of the name, 'Hilf mir Gott!' (God help me!) of a castle in the Münster valley is based on a similar event. A noble lass imprisoned in the castle was one day made the object of the vile attempts of her captor. Fleeing from his arms, she mounted the steps of the tower, and when, pursued even to this point, she saw no means of escape, saved her virtue at the risk of her life by throwing herself from the giddy height.

Unharmed, and not even stunned, she reached the ground, and her pursuer, overawed by this miracle, turned from his life of sin and iniquity and became a penitent monk in a monastery close by. 'The spot is frequently visited at nights by a spirit clad in white, and encircled by a halo of subdued light,' added the simple rustic who narrated this legend to me.

The peasant population of the country entertain a firm belief in legends of miracles worked by supernatural powers in bygone times, and it would prove highly unsatisfactory to endeavour to make a peasant realise the stupidity and incongruity of most of these miracles.

CHAPTER III.

A PEASANT'S WEDDING.

CARNIVAL, in most Continental countries a period
of general festivity, is distinguished in the secluded
Alpine valleys of Tyrol solely by the circumstance
that weddings arranged in the course of the preceding year are, if it be possible, celebrated in that
period.

Now carnival is in winter, and winter in Tyrol
is a season specially adapted for the observance
of quaint old-fashioned customs, hallowed by the
use of centuries. These striking mementoes of a
past age specially characterise a rural peasant's
wedding ; and it is in order to introduce my reader
to one of these merry-makings that I have to
request him to follow me, on a bright but uncom-
monly cold February day in 1875, to the village of
Brandenberg, a little Alpine hamlet in the valley
of the same name.

Though exceedingly heavy falls of snow had made the narrow bridle-path leading from the broad Inn valley to Brandenberg almost impassable, I had faithfully promised to so many of the frugal inhabitants of that vale to honour the wedding of a charming young peasant girl with a special protégé of mine, that I was determined to surmount all difficulties, and prove myself a man of my word.

Where in summer it would have required but a two hours' walk to reach my goal, now, in the depth of winter, it was a seven hours' battle with snow, that covered the ground to a depth of three, and in many places of four and five feet, before I found myself in the roomy inn of the village. Countless outstretched hands, brawny and muscular, small and plump, clean and dirty, were immediately stretched out to greet me. As it was Sunday, and the eve of the wedding-day, the 'Gaststube,' or bar-room, was crowded with young and old, fair and ugly Brandenbergers. My arrival, and a few minutes' conversation with my old patron, the 'Herr Vicar,' the priest, in which I sought his permission for a few hours' dancing—it is usually not the custom to dance on the eve of a wedding-day—very soon put the musicians into requisition. A couple of florins

(about four shillings) for the evening's music brought
a broad grin of satisfaction on the honest faces of
the three ' Musiker,' consisting of a flute, a trombone, and a guitar-player.

Repairing to the dancing-chamber, a narrow
room about thirty-five feet in length, I was immediately surrounded by a group of young fellows,
offering me, as a mark of courtesy, their bright-
eyed lasses. Choice was not difficult, and the next
minute I was dancing the ' pas seule,' that is, one
dance round the room, while the other couples line
the wall, and fall in at its termination.

The striking character of the national dances of
the Tyrolese calls for a few words of description.

In Brandenberg, and in some other valleys, the
male dancer encircles the waist of his partner with
both arms, while she, standing up as closely as possible, embraces him with both arms round his neck.
A peculiar and ungraceful shuffling motion is the
necessary result, and were it not for the frequent
intervals of separate dancing, the dance would be
ungainly in the extreme.

For the first minutes of every dance the motion
of the whole group is slow, and the floor trembles
beneath the heavy tramp of the strapping fellows
with immensely heavy ironshod shoes.

F

All of a sudden the music changes, and the whole aspect of the room is changed with it.

The man, letting go his partner, commences a series of capers and jumps and gymnastic evolutions, displaying an agility very remarkable, and quite unlooked for in their heavy, solidly-knit frames.

Various as these movements are, I will endeavour to describe the most striking. One of the commonest is to throw oneself on one's knees, fold both arms over the chest, and bend back till the back of the head, touching the floor, gives a few sounding raps on the hard boards, and then, with one powerful jerk, without touching the floor with the hands, to regain one's erect position.

In another the man kneels down, and with his bare knees beats a sounding rat-ta-ta-ta on the floor, and then, with one agile bound he has regained his feet.

I have tried innumerable times to imitate some of these figures; but, although I am a fair gymnast, I seldom succeed with any but the easiest.

To touch the floor with the back of the head only, with arms folded over the chest, the knees resting on the ground, is a feat which many an athlete of repute could not imitate save by long practice.

To jump high up in the air and come down upon the knees with the full force is very common.

All these capers, jumps, and evolutions are accompanied by loud shrill whistling and peculiar smacking sounds of the lips and tongue, in imitation of those emitted by the blackcock and capercaillie. Indeed many of their movements too are performed with a view to outdo the capers and circling jumps and spinning motion performed by these love-sick birds of the mountains.

The accompanying sounding slaps on the muscular thighs and on the iron-shod soles of the heavy shoes by the brawny horny hands of these fellows, the crowing, loud shouts, snatches of songs, intermingled with shrill whistling, ferocious stamping on the ground with the greatest possible force, create a din and a roar of which only they who have heard it can form any conception.

The floor rocks, the wooden beams of the ceiling tremble, the windows—if there are any—clatter as if an earthquake were shaking the very foundations of the house.

The pushing and crushing before the separation

of the couples has occurred and the whole company is yet dancing the valse, in a fashion more or less akin to the one seen in our own ballrooms, are often terrible, and the bumps against the wall or doorway are generally of huge force; but nobody shows any ill-feelings or anger, be the push ever so hard or the heavy tramp on the foot ever so painful. All is mirth, gay and rollicking fun. Now and then young fellows from the neighbouring valleys visit a ballroom for the express purpose of creating a disturbance, ending in a fight, often of alarming dimensions, if the natives are not in sufficient force to eject the rioters from the precincts of the house.

I once had the luck to get mixed up in one of these affrays. Even the musicians were drawn in, and one of them, I remember well, distinguished himself by dealing heavy blows with his brass trombone, leaving it at the termination of the disturbance a useless, misshapen mass of metal.

While the male dancer performs these odd antics, his partner, holding her short but ample skirts with both hands, continues to dance in a circling motion round him, smiling approvingly the madder and higher he jumps, or the more difficult his gymnastic evolutions.

LIFTING THE DANCER.

In Brandenberg, and one or two other Tyrolese valleys which boast of a particularly muscular fair sex, the girl at the conclusion of her swain's fantastical jumps catches hold of him by his braces and hoists him up bodily (aided of course by a corresponding jerky action of her partner), and while he, balancing himself with both hands on her shoulders, treads the ceiling of the low room to the tune of the music, she continues her dance round the room, displaying a strength and power that can only be appreciated if one has seen the strapping six-foot fellows that are thus handled by their fair partners. If many dancers crowd the room—more or less confined, if it be not a large barn—this practice is fraught with some danger, as of course when swinging himself down the dancer very frequently pitches upon some unfortunate couple who may at that moment be close to the spot where this singular gymnastic dance is about to terminate. This figure affords, of course, a very striking sight, and though there are rarely more than four or five men 'hoisted' at one time (not every one of the girls has the power, nor every dancer the requisite agility), it serves, taken as a whole, to increase the remarkable features of a 'Tanzboden,' or dancing-room, in the remote valleys of the country.

It is a somewhat erroneous impression that there exists a dance called ' Schuhblatteln,' or shoe-slapping. The term denotes merely that movement—introduced into the valse, polka, and any other of the few dances these people know—in which the male dancer strikes the soles of his shoes and his thighs with the outspread palm of his hand, accompanying this movement with the antics and the sounds I have described. Those that are unable to do this continue the round dance.

In many of the valleys the girls are passionately fond of smoking, and it is an odd sight to see many of the comely lasses pace it with a blazing cigar or pipe between their chubby lips. It is quite consonant with the etiquette of one of these rustic ball-rooms to smoke while dancing ; in fact the man who can perform any agile feat while smoking increases thereby his reputation for agility.

To place one's hat on the head of one's fair partner is synonymous with the declaration ' Thou art mine,' and beware of danger if the girl has allowed this distinction, having at the same time another swain.   Of course a native will not commit himself in this way before he is quite certain of his

A DIFFICULT FIGURE IN DANCING.

case, or if he has not the express desire to call his
rival out to fight ; but strangers, or such as may be
unacquainted with this odd custom, are not infrequently entrapped. I have seen several strangers
and tourists very roughly handled indeed by the
enraged rivals—in fact the majority of fights
among the hotheaded young fellows of a village are
caused by quarrels originating on the 'Tanzboden.'
Jealousy is in the Highlands of Tyrol no less a
feature of ardent youth than in the most civilised
country of the world ; the only difference between
the manner in which these differences are settled
being that in the former the fist, the teeth, and
unfortunately also the knife, play a conspicuous
rôle.ʃ .

I have actually witnessed only two fights that
terminated fatally, one on the frontier of Bavaria,
the other near Schwaz, in the Inn valley. In
both instances the knife was used, and the victim
was in each case the stronger of the two combatants, as fine specimens of stalwart youthful
manhood as one could see.

In the Highlands of Bavaria, as I have said
once before, the use of the knife is far more prevalent than in Tyrol, and I have known as many
as three young fellows fall its victims in one village

in one year. These knives are worn in a small
sheath sticking in a separate pocket in the leather
trousers, and as the handle protrudes it is a dan-
gerously handy weapon, though the blade commonly
does not exceed four inches in length. It is not
very long since the use of knives was prohibited by
law, and anyone carrying one was fined. This
salutary measure, however, did not long remain in
force, and the abuses of the knife are now in
Bavaria as frequent as ever.

Returning to our ballroom, we find that the
dances are short and follow each other closely, the
interval between each being filled up by a
' Schnaddahüpfler '—a short song, or rather series of
rhymes, expressing sentiments either of defiance or
derision destined for some rival's ear. It is sung
by one of the dancers, standing in front of the
slightly raised platform upon which the musicians
are seated; his girl stands at his side, generally with
cast-down eyes, and profuse blushes mantling her
cheeks. It is marvellous with what rapidity the
object of the affront or scoff will compose his reply,
replete with imputations of like or worse kind, and
in this manner two rival bards will continue for a
considerable length of time to take turns in casting
impromptu slander or scornful contempt at each

SINGING THE SCHNADDAHÜPFLER.

other. The girl, if there is no refrain to her swain's off-hand poem in which she can join, has to remain silent ; the preoccupation of the poet's mind while raking together those incidents of his rival's life which he fancies he can turn to account, and the mental labour of composing while dancing excluding very naturally the possibility of repeating the brand-new 'Schnaddahüpfler' to his partner in the five or six minutes each dance lasts. Love, of course, furnishes by far the greater portion of subjects for this modern 'troubadouring.'

A girl changing lovers, or refusing the hand of an ardent wooer, will be the welcome subject of scores of 'Schnaddahüpfler' at the next dance or wedding ; and though they are generally of a very dubious morality, these songs furnish a capital illustration of that poetic vein which marks the inhabitants of most mountainous countries, and the Tyrolese pre-eminently.

Not every young fellow ventures to fling one of these daring compositions at the head of his rival. Want of skill, or the fear of giving out after the first or second song, obliges him to be satisfied with one of the usual national lays, in which his girl, and very frequently sundry other voices, join.

At twelve o'clock the priest, carrying a huge

stable-lantern in his hand, entered the room and
ordered the music to cease. Retiring in a body
down to the bar-room, we awaited the departure of
the conscientious guardian of order ; and as soon as
his back was turned out came a 'Zither' and a 'Hackbrettel,' and five seconds later several couples were
pacing it to the charming tune of a genuine 'Landler.'
'Zither' and 'Hackbrettel' are two instruments unknown in England, and though the first may have
often been seen by tourists in the hands of Tyrolese,
the latter is much more rarely met with. Rows of
small oblong pieces of a particular kind of wood
are fixed on plaits of straw. The pieces of wood,
being of different length or shape, emit different
sounds when struck with a small wooden mallet, of
which the player holds one in each hand. Though
this instrument is very primitive and never can
rival the 'Zither'—in my opinion the most charming musical instrument existing—it does very well
for dancing purposes, and hundreds of times have
the two little hammers been in motion the better
part of a night, while I and two or three natives
were 'kicking up our heels,' making the barn or
the low-roofed bar-room resound with our vigorous
'Schubblatteln.' In this instance, as both instruments were in use, the tunes followed each other

with rapidity, and making us very thirsty, increased our beer-consuming powers to an astonishing extent.

At four o'clock we separated, each dancer accompanying his girl home—a precaution in this instance at least necessary, as fresh snow had fallen, and some of the girls had come a good distance.

Four hours' sleep in a bed—for a wonder comfortable, and not more than about eighteen inches too short—was a welcome refresher, and as I well knew the next night would be a sleepless one, I was glad to get at least that rest.

Repairing to the church at a few minutes before nine, I was just in time to see the two 'happy' couples enter the edifice. I say 'two' couples, for in this instance the ceremony was a double one, the parents of the bridegroom celebrating their golden wedding the very same day their son was married. The old couple, having the precedence, were led to the altar, a wreath was placed on the old lady's head, and the whole marriage ceremony gone through as it had been just fifty years before. After the two old people had been duly and solemnly re-wedded for the rest of their days, the young couple were led up to the priest standing on the steps of the altar. There is nothing very strik-

ing to us in the marriage ceremony of the Catholic
Church, so we will accompany the whole festive
party back to the inn, where a substantial meal was
awaiting them.   On leaving the church a bunch of
artificial flowers adorned with gold and silver tinsel
was presented to each of the 'guests,' or persons
invited to partake of the meals at the table of the
bride and bridegroom.   A 'huge specimen placed
by fair hands on my hat corroborated my fears
that I should have to share their meal, in lieu of
taking part at the shooting-match that was then
just about to commence.   A refusal on my part to
'dine' with the rest of the guests would have been
considered the height of rudeness or the result of
great pride, and as I did not wish to incur either of
these reproaches, I had to make the best of it, and
accept the seat of honour between the bride and
the 'Herr Vicar,' the priest.   My late breakfast had
reduced my capabilities of partaking of a 10 o'clock
forenoon dinner to a minimum, and enabled me all
the better to watch the feats of eating accomplished
around me on all sides.   Meats cooked in various
manners, in all of which, however, fat and grease
predominated, were the chief features of that early
dinner ; and even considering that these frugal
people rarely touch meat more than twice or three

times a year, their appetites for this delicacy were
amazing. The last dish consisted of huge cuts of
bacon swimming in a sea of molten butter, and the
hearty way this 'plat' was attacked could not fail
to increase the astonishment of an observer unac-
customed to appetites *à la* Brandenberg. Dinner
lasted three hours, and finally, after drinking the
health of the old and the young couple in numerous
glasses of wine, the party rose and made their way
to the dancing-room, where music and dancing had
been going on for three hours already, for the
benefit of those who had not been invited to dinner.
After looking on for a few minutes and applauding
the two old people's performance in a steady valse,
I retired, eager to join the rifle-match.

To the mind of a Tyrolese, the shooting-match
is by far the most important feature of any fête,
wedding, or feast-day that may have charmed him
from his cottage. Rain, wind, hail, thunder, cold,
or snow are incapable of keeping him at home when
he knows that at the next village or lonely country
inn a rifle-match is going on.

In this instance the innkeeper had arranged
the match : two 'running-stags' and two fixed targets
had been placed in the rifle-range, and the markers
at each target paid by him. He had even gone

further in honour of the occasion, and had given three prizes, consisting of silver florins sewn on large bright-coloured handkerchiefs. The priest had added another prize, and a citizen from the next townlet had sent a huge pipe, while another had presented a new rifle. Adding to these prizes the few silver florin-pieces with which I had provided myself for this occasion, I took my stand in the little shed, open on all sides, from whence the competitors fired.

My hand being still rather shaky from the wine at dinner, I confined myself at first to the fixed targets at 200 yards, presenting a bullseye six inches in diameter, provided with three rings, each an inch apart. The centre, a pin's head, counts five, the first ring, measuring two inches in diameter, counts three ; the next, four inches in diameter, two ; and the last ring in the bullseye, only one point. The white space round the bullseye is not subdivided into rings, as any shot striking blank counts nothing. Thus it will be seen that a man who cannot hit every time the No. 1 ring at least, or, in other words, who cannot pierce at 200 yards a saucer measuring six inches in diameter, has very little chance of winning a prize at a Tyrolese shooting-match. In the larger valleys, where the same attention is not

A GOOD SHOT.

given to rifle-practice, a stranger would have a
better chance ; but in the more secluded glens, where
the rifle is constantly in the hands of a man, he
must be indeed a good shot to get even a minor
prize.

An hour's practice steadied my nerves, and
I changed my position to the next partition of the
shed, set apart for the marksmen firing at the stag.
The ' running stag ' consists of the wooden figure
of a stag rigged up by means of a huge pendulum
in such a manner that, when loosened, it would dart
across an open space eight feet in width, between
tall and dense bushes. The pace at which this imitation stag travelled was about equal to that of a
living specimen in full flight. A bullseye, painted
on the ' Blatt '-region of the heart, had to be hit in the
same way as a fixed target, but of course this was a
hundred times more difficult, considering the rapid
movement of the mark ; and yet there were three or
four men present who had, out of six shots, hit five
times the bullseye—a marvellous feat, seeming well-
nigh incredible, as, at a distance of 180 yards, you
saw the stag flash past you. One of the stags was
for practice ; the other was, however, the mark
upon which nearly all the prizes were staked. A large
number of competitors being present, it was found

necessary to restrict each man to six shots at the
'grand count,' and, fortunately for me, I determined
to shoot my six shots that day, and not keep any over
for the next—the match was extended over both
days—as I dreaded 'wild' shooting, after a long
night of dancing and drinking. The sequel proved
that I had done very wisely, as all those men who
had not followed this precautionary measure shot in
such bad form the next day, that, at the termination of the match, I pulled off sixth, with a prize.

After firing my allotment I was glad to get back
into the house, as loading and shooting at a tempera-
ture of 4° Fahr. were rather uninviting occupations.
I daresay many of my readers would have been
amazed to see these men, with bare knees and open
shirt, and in many instances even without their
coats, just as they came out from dancing in the
heated atmosphere to fire a few shots, stand there
for an hour, and hardly remark that 'To-day it is a
bit cold.'

Dancing, which had commenced at ten o'clock
in the morning, was now at its height, and was kept
up without intermission till six o'clock, when supper was announced. At the morning dinner the
relatives and next friends only, not mentioning myself, had been invited. Now everybody present,

and there were considerably over 250 people, ate
and drank at the expense of the 'happy couple.'
Huge long tables with benches on both sides were
fixed wherever there was room, and the dishes, consisting of ' Knödel,' huge balls of cooked dough, with
small pieces of fat bacon, and 'Geselchtes,' a sort
of smoked pork boiled in fat rather than water,
were placed in huge bowls, as large as a moderate
foot-pan, on each table. Those who had no plates
helped themselves direct from the dishes, while large
stone jugs filled with beer, or, if the marriage is
' rich,' as they say, with wine, passed from mouth to
mouth. At our table, where the same company
assembled as in the morning, we had a repetition of
the 'dinner' dishes, and the long interval had given
me the necessary zest to enjoy the rich viands. The
din and roar throughout the house was something
terrific. Here a man, elated by his happy shot right
in the centre of the stag's bullseye, was singing a
' Schnaddahüpfler,' in which he was deriding an unlucky companion who had lost two Mass wine—about
three quarts—in a· bet on that shot; there a man
had recommenced an old quarrel with his *vis-à-vis*
about a certain chamois which both swore they had
hit, and still there was only one hole in the carcase.
In one corner a man was bawling for more drink

G

while in the opposite one two young fellows, stretched across a table, were endeavouring to settle the question of their relative muscular strength by a game of 'Fingerhackeln'[1]—their two lasses lighting their pipes with one match—and vieing to outdo each other in producing the most dense clouds of vile tobacco smoke.

Though mirth and gaiety was at its height, and wherever one looked laughing faces might be seen, there was no drunkenness among the two or three hundred guests.

Supper lasted for more than two hours. Fresh pans of 'Knödel' and huge platters of meat were for ever appearing and their contents disappearing with a rapidity most wonderful to behold. My neighbour to the right, the brother of the bride,

---

[1] The game of 'Fingerhackeln,'—interlocking of fingers, literally translated—affords one of the most amusing sights possible. The two competitors, seated opposite each other at a table, stretch their right arms across, and putting the middle finger into the shape of a hook, intwine it with that of their rival ; they then commence pulling, the object being to pull the antagonist right across the table on to the floor on the other side. Practice with a well-developed biceps frequently enables a smaller and weaker man to 'pull' his heavier antagonist in this manner. The most impossible positions of the human body, yells of despair, or growling curses and much laughter, are the invariable features of this game. In the Bavarian Highlands, where it is very frequently practised, it is considered quite an art, to be a proficient in which is equivalent to a goodly supply of beer and schnapps.

THE 'EHRENGANG.'

whose capacities in the way of 'Knödels' and 'Speck' I had watched at the morning meal, fairly outdid himself in the evening. To my certain knowledge fourteen of the former, measuring each at the very least three inches in diameter, fell by his hand, not to mention sundry hunches of the very fattest bacon ; and it was not astonishing that at the termination of his repast his head sank on his breast, his eyelids drooped, and five minutes later he was fast asleep, with his shaggy head resting on the festive board.

At about half-past nine, when most of the people had left for the dancing-rooms—a second room had been emptied of chairs and tables and devoted to dancing—the 'Ehrengang,' an institution of great antiquity, in use as early as the 14th century, began.

It consists of the presentation of money to the newly-married couple by each person, be it man, woman, or child, present at the wedding.

The chief table, where the couple had sat during supper, being cleared, a large brass or pewter dish, covered by a clean napkin, is placed at the head in front of the godmother of the bride — the mother is rigorously excluded from being present at any part of her daughter's wed-

ding. At the side of the former sits an uncle or
brother of the bride, a sheet of paper before him,
and a pencil in his hand. The gift of each guest
has to consist of at least two florins (about four
shillings), one florin being a present, the second one
is supposed to pay for the supper. Those who are
present at both meals are expected to give at
least three florins, while those who come in later
and have no share in the eating and drinking give
one florin. The money is placed in the hands of
the godmother, and is hidden by her underneath
the napkin, while her neighbour scribe notes down
the name of the donor and amount of his gift,
a proceeding which, though somewhat business-like
and odd, arises from the reciprocal custom, that when
the giver marries he expects the exact amount of
money from the bridegroom that he had given
at the occasion of the latter's wedding.

The bride and her affianced stand a little
apart from the table, she with an ever-full wineglass in her hand, he at the side of a gigantic
basket filled with huge buns of coarse flour, and
unpalatably greasy. As each guest emerges from
the crowd hovering round the ' pay-table,' the bride
presents the full wineglass, the bridegroom a bun ;
the former is drunk off to the health and prosperity

of the couple, the latter forthwith disappears in the coat or dress-pocket of the well-wisher, to be hoarded up for the next Sunday cup of coffee, or any other propitious occasion.

I was highly amused in watching the various expressions of the guests' physiognomies as they tendered their hard-earned florins to the steady matron, who just bowed her head in a stately manner as each individual pressed the two or three pieces of crumpled paper or silver florins into her hand. Now and again, when a 'fiver' made its appearance, a smile of welcome would hover round her lips; but never a 'thank-you,' or other expression of gratitude passed her lips. As the money is not hers, the thanking is left to the rightful owners, the happy couple.

No less amused would a stranger be to watch the solicitude with which the elderly female relations of the couple collect in the ample folds of clean napkins the pieces of meat, bacon, or pastry that have remained in the dishes.

Neatly packed up, they are carefully carried home, and furnish a Sunday dinner; or, if they happen to be of an imperishable nature, they are hoarded up for years as mementos of the fête.

In other parts of Tyrol presents in the shape

of furniture, such as a bed, a chest, or a table, are given, and though such gifts as these are commonly restricted to relatives of the couple, the same law of returning, at the proper occasion, exactly the same description of 'cadeau' holds good also in these instances.

A much more singular custom in the way of wedding presents is to be met with in several of the remotest Tyrolese valleys, the presentation of a cradle to the bride by each one of her discarded lovers.

At the wedding of a rustic belle, who for a series of years has held court in her summer palace, the Alp-hut, and who can boast of a whole train of ardent admirers, frequently five, six, and seven cradles, of the very roughest construction, are found in front of the house door on the morning after the wedding.

Very often it happens that just those girls who have enjoyed life to the utmost ultimately marry some man much older than themselves, who can offer them what most of their lovers could not, a house and home ; and though it may not exactly be conducive to the serene conjugal happiness of the husband to find, on awakening on the morning after his wedding, his doorway blocked up with these tangible proofs of his wife's *faux pas*, they tend no doubt to set at rest any doubts he may have entertained as to their exact number.

The 'Ehrentanz,' or the dance of honour, takes place immediately after the last guest has presented his gift. This is the solemn dance of the bride and bridegroom, the nearest of her relations, and any guest whom the bridegroom desires to honour and distinguish. All the rest of the dancers line the wall, while the host of the inn and his wife stand near to the musicians. As each couple, slowly waltzing round the room, pass the host, a full glass of wine is presented to the man, who has to present it to his partner, and only after she has drunk of it may he drain the glass. Upon the brother of the bride, or, if she has none, upon the bridegroom's, devolves the duty of singing a short 'rhyme' in praise of the occasion after each of his rounds; and now comes the most comical feature of the whole. If the bridegroom has been a gay Lothario in his day, or the bride a little too fond of her male admirers, or if, worst of all, there are any tangible proofs of her former misconduct, any one of the dancers lining the wall can stand forth, and in a gay rhyme accuse him or her of any incidents that are of a questionable character.

To these the brother, the champion for both bride and bridegroom, has to answer, and if possible retaliate with some severe cut. In Brandenberg this custom is not so generally observed as in several other valleys ; I have seen as many as fifteen and twenty of these public accusers tell tales of former sins. As they are invariably of a highly questionable character, I must refrain from giving instances.

For a rejected lover, or one that has been thrown overboard in lieu of a richer or handsomer one, this is obviously the best opportunity possible for revenging himself; and very frequently scenes of former love come upon the *tapis* that seem to civilised ears, to say the least, unseemly.

After the ' Ehrentanz ' the newly-married couple depart, and the musicians, whom thus far they had paid, are now entirely dependent upon the public. True, not quite so entirely as one might suppose, for if the receipts do not come up to their standard, they begin to scratch the fiddle, and display in other ways their contempt for the close-fisted public.

The way in which they are paid by the dancers is singular. A plate is put in front of the musicians, and after every dance one or the other of the dancers is expected to accompany his ' Schnad-

dahüpfler' song with a ten or twenty kreutzer piece
(about twopence or fourpence). After the 'Ehrentanz' the dancers settled down to real good earnest
work, to be kept up the whole night. Merrier and
merrier got the crowd, and oftener and oftener did
the glowing couples disappear to quench their thirst
in quarts of beer or gills of 'Schnapps,' spirituous
liquor.

A novel and certainly dangerous way of cooling
one's glowing face and throbbing heart is put into
practice by these hardy fellows. Coat and waist-
coat have long since been discarded as too hot,
and so in their shirt-sleeves, accompanied by their
partners, they adjourn to the well in the courtyard. While he breaks off the long icicles that
crest the spout, the lass lays hold of the pump
handle, and in the icy cold water that spurts forth
he bathes face, neck, and chest! And yet consumption or any complaint of the chest is, if not
quite unknown, of very rare occurrence in these
valleys.

Dancing ceased at six o'clock in the morning,
for the tolling church-bell announced early service in
honour of the saint whose 'day' it happened to be.

At seven o'clock when service was over we were
again at it with fresh vigour, obtained, in my case at

least, in the shape of a very solid breakfast. An
hour later shooting in the range commenced ; but
on trying my luck, when I finally got tired of
dancing, I found that a night's 'spree' does not
tend to steady one's hand.   I gave it up as a bad
job after firing some ten or twelve rounds.

Recommencing dancing with a batch of fresh
fair dancers—who had not been up the whole night
—the fifteen or twenty young fellows, including
myself, who had determined to hold out as long as
there was a nail in our shoes, were animated with
fresh strength.   We kept it up, with an hour's
intermission for dinner, till 6 o'clock that evening,
or, in other words, we had accomplished the feat
of dancing more than thirty-two hours, with the
sole break of the four hours that had been given
up to sleep the first night.

After indulging in a hearty supper we commenced our preparations for our start homewards.
Three young fellows, natives of a village close to
my home, had decided to accompany me that night
rather than to stop the night at the inn and return
next morning.

Provided with huge bundles of pine torches
and ·a bottle of 'Schnapps,' we started at about
eight o'clock that evening.

Heavy falls of snow had obliterated every trace of the steps that had been imprinted in the deep snow the previous day, thereby materially increasing the difficulties of our task.

Though we had, all four of us, broad snow-hoops on our feet, we sank far beyond our knees in the yielding mass of snow.

Had I not been so fatigued by my uninterrupted dancing the two previous days, our march home would have been a pleasing and interesting finish to my mid-winter expedition to Brandenberg.

Silently we pushed on for many hours. The glare of the torches, the mysterious silence of nature under a heavy pall of snow, the ghostlike appearance of the trees, the odd and fantastical shadows on the white background, and finally the dull thud and roar now and again when a tree, giving way under the weight resting on every portion of it, snapped asunder, were all features of my nocturnal return home from a peasant's wedding.

In many of the larger valleys, as for instance the Unter-Innthal, Zillerthal, and Brixenthal, which, as the German phrase has it, 'are licked by civilisation,' the old wedding customs have of late years, to a great extent at least, been done away with.

In some instances innovations in these quaint and pleasing relics of bygone ages were a source of contention for that part of the population who, though the shriek of the locomotive was within earshot, were not ashamed to continue to do as their forefathers did.

Several years ago an instance of the general unpopularity in which the modernised wedding customs were held came under my immediate notice. A wealthy young 'Wirth' — who had been for several years in Munich and Vienna, imbibing there a predilection for town manners and habits—had his wedding with a damsel of his native townlet conducted strictly on 'town principles,' inviting only a limited number of guests, doing away with the usual public dancing, and in fact turning the usual merrymakings at a rural wedding into the torturously wearisome ceremony prescribed by the rigorous code of civilisation. The young fellows and fair lasses of his native townlet took this remodelling of time-honoured customs, and particularly the fact that they were deprived of their dance, greatly amiss. Not content with showing their dissatisfaction in various ways, they determined to carry out the bright idea, proposed by one of them, of arranging a mock-wedding on the very day and in the very inn selected by the object of their wrath for the solemnisation of his marriage.

The indignation and wrath of the pompous bridegroom can be fancied when he perceived an exact counterpart of his own ceremony, going into every detail, such as the same number of carriages, the same number of ' Böller' shots—small cannon— take place. Short of the actual marriage scene in the church, the comic farce was an exact copy of the genuine ceremony and the subsequent festivities. The roomy Wirthshaus, the site of both wedding dinners, was divided into two antagonistic strongholds, the genuine guests occupying the rooms on the ground floor, the sham ones disporting themselves in the upper apartments. A band of music having been provided by the latter, dancing commenced shortly after dinner; the male guests of the bridegroom, numbering about a fourth of their uproariously gay enemies, and being obliged, therefore, in view of the heavy odds that would be brought to bear against them if any quarrel arose, to keep very quiet, had not only to pocket the insult of the whole proceeding, but actually were constrained to stand by and witness their sisters, daughters, or sweethearts carried off to the dancing-room by their rivals.

The sham bride, a dressed-up man, brought the matter to a head by entering the room tenanted by the bridegroom's party, and going up to him knocked the hat off his head and picking it up placed it on his own. I have said what the act of placing one's hat on a girl's head means. The bride, bursting into tears at this further indignity, upbraided her affianced for his conduct. The latter, stung to the quick by the whole affair, was just about laying hands on the fiend in woman's shape, when a body of gendarmes—the rural police —entered the room and put a stop to any further disturbance. The host, well aware that a fight on a grand scale would very probably be the finish up of this whole farce, had despatched a messenger on horseback, at an early hour in the morning, to the next town to fetch a body of these peacemakers. Their arrival in the evening occurred, as we have seen, in the nick of time ; a few minutes later and they would have found the whole house a scene of fierce fighting, on a scale rendering even the inter-vention of twenty or thirty gendarmes but little use. As it was, three gendarmes, posted at the foot of the stairs, cut off all communication between the two hostile parties, and were able to keep the peace for the rest of the night.

CHAPTER IV.

THE WOODCUTTER.

THOUGH I have not laid special stress on the fact that Tyrol possesses certain characteristics not to be met with in other parts of civilized Europe, the reader will no doubt have gathered this from the remarks in the preceding chapter.

The survival of an ancient type is in no class of the population so apparent as in the fraternity of the woodcutters.

Cut off from the world, working in solitude amid the grandest of Alpine scenery, rough and uncouth in their exterior, inured to every danger, and hardy to quite an amazing degree, the ' Holz-hacker' affords a most interesting study not only for the artist, but also for those who delight in laying bare the vein of quaint originality mixed up with the other characteristics of a people untouched by that species of civilization which follows in the wake of tourists.

The immense tracts of forest which are still to be found in the northern and centre districts of Tyrol, and which afford the staple resources of those parts, are generally speaking the property of the Crown.

A large number of men are employed by Government in felling the timber, in cultivating new plantations, and in keeping in repair the huge wood-drifts which are established in these parts.

From 3,000 to 4,000 men thus find sustenance in connexion with the 'Forstwesen,' or management of the forests, in Tyrol.

These labourers are generally natives of neighbouring valleys, and in most cases they are younger sons of peasants—farmers who own the land they till—whose miniature homestead, consisting perhaps of a few acres of the very poorest soil, or a patch of meadow sufficient to keep three or four cows, proves inadequate to sustain an increasing family. The eldest son usually remains with the father, nominally inheriting the whole property at his death.

I say nominally, as, by virtue of the old laws of

inheritance passed in the end of the last century, a division of the property is inadmissible, and the happy nominal owner is not a whit better, if he be not worse, off than his brothers ; for on the death of the father a Government appraiser values the property, fixing the estimate rather higher than the real value. This sum is divided into as many equal shares as there are sons,. each of whom receives a mortgage on the property for the amount of his share.

The eldest son, in lieu of his share, takes possession of the property, and endeavours, by dint of the greatest economy and care, to pay off mortgage after mortgage. If he fails in this, or if he is a spendthrift, his children, if he has any, are doomed to be paupers, as a further division of their father's share does not take place, and the property is sold. Not infrequently the mortgagors, unwilling to let their home pass into strange hands, club together and buy it up ; or if they cannot muster a sufficient capital between them, they with one consent cancel the debt, and install as master of the concern the one who has the most knowledge of farming, and in whom they have the most confidence, or if none are willing to undertake the charge, one of their nephews.

H

The daughters of a peasant either receive a certain sum as dowry, or, if they are unmarried at their father's death, the few hundred florins which have been saved up by their parents fall to their share.

It shows well for the Tyrolese that in many of the remoter valleys the peasants date the history of their family and that of their property back for many centuries, and the old crossbows and pieces of armour, which are frequently to be found among the rubbish in the loft under the roof, tell tales of former bondage and serfdom to the person of the next knight or baron.

Returning to the lot of the younger sons, I must here mention that the choice of their profession depends entirely upon the customs which are prevalent in their valley. Some few valleys furnish the wandering hawkers of carpets and manufactures of plaited straw that turn up at large fairs throughout Europe ; and I am speaking from experience when I say that no capital in Europe is without a few of them. The inhabitants of some glens have acquired the art of carving figures in wood ; other valleys produce hawkers of gloves and articles of chamois leather. While one Alpine glen is celebrated for its ' Kirschwasser,' a spirituous liquor distilled from cherries, another is renowned for a particular kind of cheese.

Three or four centuries ago Tyrol was the richest mining country in the world ; but now most of the prolific gold, silver, and copper mines are exhausted, and only two or three valleys contain mines which pay.

In each of the valleys I have enumerated the whole population, save perhaps the peasant-farmer, is interested in the special branch of occupation which is its distinctive feature, and which tends in a more or less injurious manner to make the people acquainted with the outer world, its ways and its habits, thereby occasioning that gradual loss of the ancient typical customs, the partial survival of which I pointed out in my introductory remarks as one of the attractive characteristics of Tyrol. In those valleys where forests form the chief resource of the inhabitants the results of contact with the outer world do not appear. The occupation of a woodcutter, the scene of his thrifty labour, and his own predilection take him far out of the way of railways and tourists.

For seven or eight months of the year he is out among the mountains ; the rest of the year, when the huge quantity of snow makes outdoor occupation impossible, he retreats to his home, now doubly and trebly secure from any attempt of a tourist to push his way into these nooks and corners of the Eastern Alps.

Many of these hardy fellows have never seen a railway, and Bismarck and Moltke might conquer the universe without their knowing anything of it.

Have any of my readers ever been asked—as I have—if London is a village in Welsch-Tyrol (the southern part, where Italian is spoken), or if England is a town in Bavaria? Borrowing the phrase from our American cousins, I venture to say, 'I guess not!'

After this digression, which was needed to place the character of the woodcutter in the proper light, let us return once more to his occupation. The youngest and strongest men among the three or four thousand who in one way or the other find employment in connexion with the forests are the fellers of timber.

Their vocation is one in which dangers, arising from the most varied causes, and from exposure to all the inclemencies of a rough Alpine climate, make an iron constitution, a clear head, and powerful body indispensable. What would my reader, be he a retired backwoodsman or not, think of living from March or April till November on a mountain slope in the close proximity, perhaps, of vast snowfields, and rarely at a lower altitude than 5,000 feet over the level of the sea, in a hovel, the roof and sides of which are of the thin and porous bark of the pine-tree? Yet thus they pass the summer months, and more content and cheerful fellows than they are it would be impossible to find.

The dangers which beset their rugged path are numerous. They arise either from their own recklessness, from avalanches, landslips, or from elementary causes, such as lightning and the devastating effects of waterspouts.

Tourists are often astonished at the wonderful number of sacred pictures, shrines, and votive tablets which line the highways and byways of the country. In nine cases out of ten they simply commemorate a woodcutter's violent death, or some other fatal accident which has taken place on or near the spot. In the larger valleys these votive tablets are generally some fearful specimen of the local stonemason's brush, who in his leisure hours turns artist, and 'paints' sacred subjects to order. In the more remote valleys similar fatal occurrences are commemorated by pictures representing the accident itself.

Underneath the painting a few lines acquaint the passer-by with the name of the unfortunate victim, and add a request to pray a couple of 'Vater unser' (Paternosters), for the benefit of his soul. The wording of these epitaphs is, if it were possible, even more ludicrous than the style of the picture which heads them. Two or three samples, literally translated, will corroborate this.

In the first we see a falling tree, under which, spread-eagle fashion, a man lies. The epitaph runs: 'Johann Lemberger, aged 52¾ years. This upright and virtuous youth[1] (Jüngling) was squashed by a falling tree on the 11th December, 1849. Pious passers-by are implored to say three Lord's Prayers to redeem his tortured soul from the fires of purgatory.'

The second represents a woman falling down a precipice ; the epitaph runs as follows : 'On that rock yonder perished the virtuous and honoured maiden, Maria Nauders, in her twenty-second year. The kind wanderer is begged to release " two " purgatoried souls from the tortures of hell.

' This wench was with child.'

[1] Unmarried men are called ' youths ' all their lives.

A third, rather more laconic, runs :

'MICHAEL GERSTNER,
' Climbed up, fell down, and was dead.'

The picture of a man falling down from an *apple*-tree made it clear *why* the unfortunate Michael had climbed it.

A very comical picture near the ' Kaiserclause,' a large woodrift, depicts three men sitting, one behind the other, a-straddle of one large block of wood, which is in the act of being drifted down the turbulent and foaming waters. Each man has a cross over his head, and the expression of the faces is comicality itself.

This epitaph is one of the best of its kind, and shows a good deal of humour on the part of its author : ' On this spot did Johann Memmen, Christoph Müller, and Alois Hausler, on the 24th June, 1838, set out on a long and perilous journey. They hoped to find the gates of heaven open.'

Underneath this is a picture of the three men in the furnace, and below that again is written : —

' In case their journey ends in hell the pious wanderer is requested to say the rosary to save them from some of the tortures which await them.'

Were it in my power to add the orthography of the epitaphs, it would greatly heighten the effect of these primitive and curious remnants of a very ancient custom.

The reckless daring which is a prominent feature in the character of a woodcutter, is the natural result of a hardy confidence in his own powers and a long immunity from accidents, and makes him look upon the most urgent precautions dictated by his craft as needless. The felled tree falling a moment too soon, or the sharp axe glancing off from the hard frozen wood, are only too frequently the origins of votive tablets.

Drifting the wood, too, though apparently a very safe occupation, is the source of many accidents, as we have seen by the fate of the three travellers the subject of the last epitaph.

A short sketch of the opening of a drift will give my readers an idea of the sort of work which falls to the lot of these fellows.

The timber which has been felled in the course of the autumn and spring on the slopes of a valley is brought down to the waterside in May and the commencement of June. Important wood-valleys have a wood-drift of their own, erected by Government. It consists of a huge barrier of the strongest timber at the upper end of the valley, right across the drift stream. On the upper side of this structure a deep reservoir is excavated, in which large quantities of wood accumulate, thereby raising considerably the water-level. As soon as this artificial pond is filled with timber and water the ponderous iron-bound gates of the drift, thus far tightly closed, are sprung open, and with a terrific roar, making the earth around shake, the water and huge blocks of wood rush through the barrier on to their destination, frequently ten or fifteen miles further down, and close to the conflux of the drift-stream with a larger one, when the wood is caught up and piled in huge stacks. Drifts are necessarily erected only in streams in which the ordinary water-power would prove inadequate to float timber measuring from three to eighteen feet in length, and from two to five feet in diameter.

If the drifting stream takes its course through narrow gorges and defiles of walls of rock several hundred feet in height, the floating of timber calls for great exertion on the part of the men engaged in it. In these places the timber is very liable to get jammed together. In a few minutes the whole bulk of the wood, very often 2,000 or 3,000 'klafter' or 'cords,' may choke up the narrow passage in one stationary mass, while the water runs to waste, either in channels underneath the mass, or by overflowing it. When one of these 'blocks' occurs, the men have to be lowered by ropes from the brink of the chasm above ; and with saws and long poles, provided with ponderous iron hooks at one extremity, they strive to bring the whole mass into motion by sawing through the timber which has produced the block, or if this fails, by working off block after block, which latter often requires the incessant labour of months.

The dangers which attend this occupation are very obvious. If the mass should begin to move again before the men standing about in different positions on the blocks are prepared for it, and before they have regained their ropes, they arc inevitably crushed to pancakes by the bumping and crashing timber.

There are instances in which a whole party of men, numbering twelve or fifteen individuals, has perished in this manner.

Where, again, the stream covers a large surface, and is dotted here and there by huge boulders that have tumbled down the precipitous slopes of the valley, the drifted wood is sometimes caught ; or if the banks are shallow, a huge block will get stranded or shoved up high and dry by the impetuous rush of the blocks in its rear. In such cases the men have to stand up to their waists in the icycold water the livelong day, while endeavouring to push block after block back into the turbulent stream, the least inattention or carelessness on their part being followed by disastrous consequences. The fellers of the timber, on the contrary, have, by the time the drifting begins, already been for some time high up on the mountain slopes, preparing a fresh stock for next year's drift ; and if my reader will follow me on an unsuccessful chamois-stalking expedition, which brought me into a woodcutter's hovel high up on the Tyrolese Alps, he will make the acquaintance of as quaint and primitive a set of human beings as can well be met with this side of the ocean.

A thunderstorm in the High Alps is a somewhat hackneyed subject, numerous authors of Alpine literature having been caught by thunderstorms which surpassed everything of the kind hitherto known.

It was during one of these grand spectacles that I was picking my steps down a rugged and steep Alpine path, after my unsuccessful chase. A stay of three days and two nights among the peaks and grand snowfields had exhausted my provisions, and I was obliged to seek hospitable quarters in the little Alpine valley lying some five or six thousand feet below me.

Securing the lock of my rifle, and covering my ' Rücksack ' with a waterproof hood, I cared little for thunder and lightning, and the heavy downpour of rain which accompanied them.

Soon after reaching the line of vegetation my path led me through a dark and gloomy forest of huge patriarchal old pine-trees, coated with gigantic moss beards yards in length, which imparted a vivid appearance to many an oddly-shaped tree. After having walked some time down the steep slope, vaulting now and again over the prostrate form of one of these giants of the forest, I came upon a large clearing. The huge stems, like hoary monsters slain by a dwarf's hand, lay scattered about in reckless confusion, while the fresh surface of the stumps indicated that axe and saw had been but very recently at work. Proceeding down the edge of the clearing, and making mental calculations of how many thousand per cent. profit one would derive by the transmission by fairy hand of a batch of these huge trunks to any of the large timber-devouring cities in England, I perceived a few minutes

MY ENTRANCE INTO THE WOODCUTTERS' HOVEL.

later the miserable hovel of the destructive dwarfs, the wood-fellers.

A thin wreath of blue smoke curling up in spite of the rain from a hole cut in the roof convinced me that my anticipation of finding the dwelling inhabited was correct.

Well aware that no other human habitation was within a five or six hours' walk at the very least, I gladly availed myself of the hospitable ' Geh eina, Bua' (Come in, boy !)—young men up to the thirtieth year are invariably termed boys—which greeted me on showing my dripping head inside the low doorway.

Four men, all wood-cutters, were sitting round a roaring fire, and though it was hardly half-past five, they were busy preparing their evening meal, the appetizing odour of which reminded me in a most inviting manner that I had not tasted a warm dish of any kind since leaving home some three days before.

The usual questions, ' Who art thou ? ' and ' whence dost thou come ? ' having been answered by me to the satisfaction of my hosts, I had in the twinkling of an eye divested myself of my dripping coat, shoes, and stockings, and placed them as near to the fire as the arrangements of the party permitted.

I may as well mention that on such occasions I carefully refrain from playing the fine gentleman. For the question who I am and whence I come, I have suitable answers, for were they even to learn that I am not a native, but a stranger, shyness would take the place of frank, open-hearted mirth, and suspicion at the probable purpose of my presence in so outlandish a place divest a meeting of this kind of all its characteristic features ; and to make myself accurately acquainted with these characteristics had formed, to speak plainly, one of the causes of my attachment to Tyrol.

The primitive interior and exterior of this hovel call for a few words of description.   To begin with the construction of the building, which, it must be remembered, is the work of a few hours for three or four men, we first of all find four stakes driven into the ground.   They are the corners of the edifice, and, in order that the roof may receive the necessary incline, one pair of stakes are left longer than the other two, or they are of equal length, but the upper two stand on rising ground.   The tops of these four stakes are connected by stout poles, and across these rows of laths, or, if they cannot be procured, fir-branches are laid. On these again the roof, consisting of large sheets of the bark of pine-trees that have been soaked for some time in the next streamlet, are nailed with wooden pegs or weighed down by heavy stones: the sides or walls are of the same material. Woodcutters' huts are rarely more than 9 to 11 feet square, except when they are erected for permanency, and then they are log-cabins varying in their size according to the numbers which are to live in them.

The present one was not larger than 9 feet square. The fireplace, a heap of stones raised to about 2 feet from the ground, occupying the centre ; the outlet for the smoke, a square hole in the corner, opposite the low and narrow doorway, unprotected by a door of any kind ; and finally, the four slanting boards in lieu of beds, were the chief objects that struck the eye as one entered.

Each man had his haversack hanging on a peg over his board ; the latter, covered by fir-branches and a rough blanket, must have proved a somewhat hard, uncomfortable, and cold couch for six or seven months of the year.

The huge iron frying-pan, filled to the brim with ' Schmarn ' (flour, water, butter, and salt), sus-.

pended by an ingenious mechanism over the roaring
wood fire, was beginning to utter signs of welcome
import.

Plates, dishes, tables, and chairs are unknown
luxuries in one of these dwellings. The pan, placed
on a huge log measuring some three feet across the
level surface, was our plate, dish, and table in common ; the spoon, invariably carried along with the
sharp knife in a separate pocket of the owner, con-
veyed the steaming mess from the pan to the
mouth, and a small barrel, holding some eight or
ten quarts of water, with a hollow piece of wood an
inch or two in length placed near the bung-hole,
was our glass and jug.

It requires a very formidable appetite to be
able to eat any quantity of a genuine woodcutter's
' Schmarn.' Terribly greasy, it satiates with marvellous rapidity; and one can only look on with astonishment at the incredible quantities which these
men will consume. They eat it three times a day;
in fact it is their only food, save a hunch of
bread, and perhaps now and again a few slices of
bacon.

A small bag full of tea invariably forms part
of my chamois-stalking kit, and so, after the dis-
patch of our supper, I proposed to indulge in the

inestimable luxury of a panful of tea. Now to
the mind of a Tyrolese the word tea (or 'Thee') con-
veys anything but an agreeable impression. Teas are
with them the simple decoctions of herbs and leaves
of certain trees and bushes, used only for medici-
nal purposes. Thus they have a tea for coughs, a
tea for pains in the chest, another for bile, rheuma-
tism, and even, strange to say, a tea for sprained
ankles or dislocated joints ! My proposition, therefore, called forth the usual inquiry—'Wo feilts ? '—
Where is the ailing?' My explaining to them that
this was Chinese tea, and that certain nations drank
it once or twice every day of their lives, created a
general laughter, and the covert hint that no wonder the 'Städtler,' or people from towns, were such
pale-faced and spindle-shanked individuals.

Filling the pan with clean water, I readjusted
it over the fire, and looked about me for a second
vessel into which to pour the boiling water. My
inquiry to this effect brought forth a somewhat odd
'teapot.' It was a tin washhand-basin, knocked
in and beat into a hardly recognizable shape. The
traces of lard on its sides indicated very plainly to
what use it had been put, namely, for the convey-
ance of their store of this indispensable commodity.

Well cleaned with hot water, it was a capital

I

substitute for a teapot, and often I have not even
had one so serviceable.

After placing a handful of tea in a muslin bag,
expressly reserved for this purpose, and putting
the latter into the 'teapot,' I poured the boiling
water over it; a few minutes later a steaming bowl
of tea, free from the leaves, which remained in the
bag, was standing on the log.

Sweetening it with some sugar from my store,
I invited my companions, who had been watching
my proceedings with a half comical, half serious
expression of face, to partake of the 'Chinese tea.'

A few drops satisfied them, and they put down
their spoons with the hint that they were not ill.

Well knowing their tastes, I first of all drank
as much as I wanted, and then poured an ample
allowance of 'Schnapps' into the tea. This produced a great change for the better, as my hosts
informed me, and they finished the basin with
great relish. Far more, however, than the tea did
they admire my tobacco, and soon the hut was
filled with dense clouds of my birdseye (smuggled
into Austria at the cost of great trouble and
stratagem), of which, being an inveterate smoker,
I always carry a goodly store with me on expeditions of like kind.

Tea and tobacco had loosened our tongues as only those two comforts of life can do. Merry songs, gay stories of sporting exploits or serious adventures, told in a quaint, pleasing fashion, that attracts the listener in an inexplicable manner, went round, making very frequently the frail structure over our heads resound with our merry peals of laughter.

The cold night air—we were at an altitude of considerably over 6,000 feet—and the splashing of rain, that found an easy ingress through the unprotected doorway, the smoke-hole, and various clefts and holes in the sides and the roof of the hut, made me glad of my coat, while these marvellously hardy fellows, in their shirt-sleeves, open shirt-front, and short leathers, displaying limbs of truly gigantic power, and knees as scarred and scratched and mahogany-hued as one can possibly imagine, seemed as comfortable and warm in their scanty attire as if the midday sun of a summer's day were shining upon us.

Two of the four woodcutters turned out to be noted poachers, and after I had gained their confidence by means of several little knacks with which long practice has made me acquainted, they came out with some of their adventures while following that dangerous craft. They produced their rifles—hidden among the dry branches of the roof—and showed me their simple but effective mechanism. The stock, namely, could be un-screwed from the barrel, and thus the whole rifle could be carried underneath the coat or in the 'Rücksack,' without awakening suspicion in the mind of any keeper who happened to meet them. The older of the two, a man of about thirty-two, had had several very close encounters with the keepers of the neighbouring Bavarian preserves. A terrible cut, disfiguring his whole face, was one of the wounds, while the brawny back he exposed to my view to corroborate his tale bore in numerous holes the marks of a gunshot wound.

On my asking him when and how it happened he replied, with a somewhat grim smile, that he was willing to tell me the story, 'For,' he added, 'that shot,' meaning the one in his back, 'was the last one that ——(keeper) fired. Why did he miss me with his rifle? As if I cared much for these peas at a distance of more than forty yards!' The fact that many keepers carry double-barrelled guns, one barrel rifled for ball, the second for shot, explains these words. The keeper had missed the poacher with his first barrel, and instead of keeping his shot till closer quarters, had fired it when the poacher was yet some forty yards distant. The latter had turned instinctively when he saw the keeper intending to fire, and thus received the small-sized shot in his back, doing but little injury, and without preventing him from taking vengeance in too summary a manner on the person of the foe, who, I must add, had shot at him on a previous occasion.

The second poacher, my neighbour to the right, I knew by reputation.

Of gigantic build, rare power and agility, he one time succeeded in beating off three keepers. They had just left an Alp-hut in order to fetch some wood to make a fire, and had left their rifles in the inside of the châlet, when all of a sudden ' Dare-Devil Hans ' — the name by which my friend went—appeared on the scene. Perceiving that they were armed only with their ' Alpenstöcke ' and a hatchet, he placed himself with his back to the outside of the closed door of the hut, and defended himself so bravely with his alpenstock against his would-be captors that he not only injured two very severely, but actually put them to the rout, bagging their three rifles and a chamois as his legitimate spoils. Two years after his relating me this tale the poor fellow had to pay with his life for his daring raids in strange preserves. Like numbers of his brethren, he fell a victim to the hatred of his relentless foes, the keepers. Shot right through the body, he had yet sufficient strength to outstrip his pursuers, and, faint with loss of blood, he made his way to the distant Alp-hut tenanted by his girl, only to expire in her arms the following day.

To show how close temptation lay to my hosts, I may mention that they had simply to cross a sort of gorge, ascend the opposite slope, and they were within the boundaries of a royal Bavarian preserve splendidly stocked with game.

Saturday afternoon and Sunday are the woodcutters' days of recreation. The men either follow their perilous sport, or they visit their sweethearts in their solitary châlets, or they descend from their lofty perch and make their way to the verdant valley, whence, staggering under the potent influence of strong liquor, with bags filled with flour, bread, butter, and lard—their provisions for the next fortnight or three weeks -- they reascend late on Sunday night. Their wages, I may add, vary between 90 kreutzers and 1 florin 40 kreutzers (1*s.* 10*d.* to 2*s.* 10*d.*). The proceeds of poached game are generally ridiculously low, for the innkeeper who buys it knows very well how they have come by it, and the vendor has to accept quite nominal prices. Thus a roebuck fetches 2 to 3 florins (4*s.* to 6*s.*), and a chamois even less.

We retired to our couches at a late hour ; quite soon enough, however, for me to pass an uncomfortable night, wedged in between two of my strapping hosts. At half-past four we were up cooking our breakfast, and while they were buckling on their crampons—these men hardly ever work without them on their feet—I examined my rifle, intending to enjoy a stalk on my way home.

The rain was still coming down in torrents, and the rivulet, quite an insignificant watercourse the night before, was now a swollen and roaring torrent.

We were just about to set out on our different vocations when in rushed a man dripping with water. It seems that about two hours off another gang of woodcutters were at work. Their hut, built on the brink of a rivulet, had been torn away in the night, while they were sleeping, by the rushing and roaring masses of water of the rivulet, now a mighty torrent. Two of them had been injured —one rather severely, the man told us, the other but slightly. He had come to ask us to aid him and his comrade to transport the injured men to the nearest houses, where medical aid could be procured.

Of course we all were ready to accompany him, and putting our best foot foremost, we reached the scene of the disaster within an hour and a half from the time we started. Not a stick or vestige of the hut remained to indicate the spot where it had stood.

The poor fellows were in a sad plight; they had lost their provisions, bags, axes, and crampons; and though the two latter articles were subsequently recovered some considerable way down the bed of the torrent, yet their loss was for them a very severe one.

By means of a litter made of two long poles, some pine-branches, and my blanket, we transported the severely-injured man to the next house, five hours off; while his companion, who had been stunned, had recovered himself sufficiently not to require our help. He and one of his confrères remained at the scene of the disaster in order to raise another hut in a more secure spot. About noon we reached our destination, the first house of a straggling little hamlet.

The doctor, who lived in a large village some fifteen miles off, was immediately sent for, and about 10 o'clock at night he arrived, accompanied by our faithful messenger.

The injuries which the man had received were severe, but his strong constitution pulled him through ; and when, some four or five months later, I had occasion to pass through this hamlet again I was told that he had joined his mates some weeks before.

It must seem strange to readers surrounded by luxuries and comforts of every kind to hear that a patient had to wait ten hours for medical assistance. This, however, is by no means a particularly long delay in the arrival of medical aid. I have known forty-eight hours elapse after an accident before the doctor or surgeon came. In winter it is often quite impossible to cross the mountains between straggling hamlets and the next village which boasts of a doctor. That the duties of a medical man in the rural districts of Tyrol are excessively arduous — and they are shamefully underpaid by Government—we can well fancy.

In many of the villages the doctor has to leave his bed, winter and summer, at half-past three o'clock in the morning to attend to the peasants

who need his advice. They come from the surrounding heights and mountain slopes—their homes
—to attend the four o'clock early mass, and prior
to their entering the church they look in upon the
doctor, state their ailings, and then, at half-past
four, when mass is over, they fetch the medicine,
which the doctor has made up in the meanwhile.

To return to the wood-fellers: I have yet to
relate a little adventure which I once experienced
along with three of these rough, original beings.

We had been shooting in the preserves of my
companion's native village, skirting the Bavarian
frontier for many miles.   I had been unsuccessful
on both days, when at last, towards the evening of
the second one, I got a shot at a splendid stag
carrying fourteen points.   He had come up a sort
of ravine, and was just breasting the top, when my
ball entered his chest, striking it, however, in an
oblique direction.   My ball, a large one, failed
to penetrate the animal, but nevertheless brought
him down upon his knees.   The Bavarian frontier
was not more than a hundred yards off, and
should the stag succeed in regaining the use of his
limbs and crossing the frontier line, he was lost
to us, further pursuit involving great danger, on
account of the ever-watchful Bavarian keepers.

THE STAG BREASTED THE GORGE.

Hastily reloading my rifle, I made for the spot where my victim was kneeling. To reach him I had to scramble down some very precipitous cliffs, at the bottom of which a small stream ran. Intending to ford this stream at a certain point, I rushed down the cliffs. On reaching the bottom I saw that I had mistaken the site of the ford ; but it was too late to stop my headlong course, and the streamlet being too broad to be crossed by a flying leap, I and my rifle were floundering a second later in a deep hole worn in the solid rock by the action of the water.

On regaining the shore, a matter of some difficulty, owing to the smooth, polished rock that surrounded me on every side, I put aside my now useless rifle, and, armed with my knife, I hastened up the steep cliff flanking the gorge to the spot where I expected to find the stag. He was gone, and the gory track left no doubt in what direction— of course down the ravine, right into the Bavarian preserves. My mortification can be fancied ; a ' fourteener '—a rare piece of good luck—to be lost at the very moment of success. The wounded hart could not have gone far, very probably not further than a few hundred yards, and there, breaking down, would die a lingering death within a few paces of the frontier.

My three companions, attracted by my shot, soon made their appearance. To pursue the wounded stag would be certainly a very risky undertaking, and yet we could not leave the noble animal to its fate. My companions, though woodcutters, were in this instance no poachers, and entertained a wholesome dread of the sharp practices of the Bavarian keepers, who often follow their call to surrender by the sharp bang of their dreaded rifles. We decided to refrain from taking any decisive step that evening, but rather to await the morrow. By that time, we hoped, any keeper who might have been attracted to the spot by my shot would have left, leaving us free scope to pursue the wounded hart. Dawn of day found us tracing the track of the stag across the frontier down the slopes of the ridge, along the height of which ran the boundary line. We had not proceeded for more than a mile at the utmost when we came upon the stag, stretched out below the overhanging boughs of a huge pine ; he was yet living, though evidently in a dying state. The 'Knickfang' with my hunting-knife, *i.e.* the severing the spinal cord at the point where neck and back join, soon put the poor animal out of its pain. To enable the reader to understand the details of

the following incident, I must mention that the
·tree under which the wounded stag had taken re-
fuge stood in the centre of a clearing, flanked on
two sides by high bluffs, while steep precipices
hedged it in on the two other sides. We were just
preparing to brittle the noble animal, intending to
quarter it afterwards, in order to carry it off in this
way, when, without the slightest notice on the part of
our assailants, two shots were fired at us. The distance was however fortunately so great—the keepers
were ambuscaded behind some bushes on the top
of the bluffs overlooking the level clearing—that
both struck the ground some yards from our position. We did not give our foes time for a repetition of the volley, for, with sundry angry oaths, my
three companions collected their rifles and the sacks
they had laid aside, and, following in my wake, we
gained the sheltering wood, and some minutes
later our own preserves in safety. Of course the
stag was lost to us, the keepers not only obliging
us to retreat, but being rewarded for their watching
by a noble 'fourteener.'

CHAPTER V.

THE CHAMOIS AND THE CHAMOIS STALKER.

VERY frequently have I been astonished at the degree of ignorance displayed by the travelling public respecting the chamois and its habitat. In fact it would seem that in the minds of most people this animal is associated with tales of miraculous feats, intermingled with a superabundance of romance and superstition.

Let us endeavour to fathom the cause of this odd anomaly : an animal inhabiting the very centre of Europe, and yet enveloped in a veil of mystery.

The extraordinary powers of locomotion with which the chamois is gifted, and the elevated nature of its home, make its pursuit by man a difficult and dangerous task, requiring constant training from childhood, together with courage, an iron constitution, and a clear and steady eye and hand. These qualities a chamois-stalker must possess, and very naturally it is just these that remove chamois-stalking in its genuine sense from the hands of educated and scientific men to those of the hardy native, who, while willing to undergo the necessary fatigues and privations, has the muscles and heart that furnish a 'Gamsjäger.'

To a native chamois stalker—the only person, as I have shown, who has the opportunity of watching the movements and habits of that animal—the idea of watching his game with any other view than that of sport would seem supremely ridiculous.

Saussure and the late Mr. Boner are perhaps the only two persons who have described the chamois accurately and from their own experience. The Saussure of the eighteenth century found the Swiss peaks still tenanted by the fleet tribe of chamois, while Mr. Boner laid the scene of his observation and sport in the somewhat tame scenery of the Bavarian Highlands, where sport is made easy by large preserves, and the far less precipitous and dangerous nature of the sporting grounds.

While Switzerland has been effectually cleared of its former tenants by the invading hosts of tourists and travellers, Tyrol has, by dint of some judicious game-laws, managed to increase its stock to a very considerable extent.

The three largest preserves in the country, viz. the one near the Achensee, belonging to the Duke of Coburg, the one situated near Kufstein, the property of Archduke Victor, brother of the Emperor of Austria, and the preserve occupying the extreme end of the Zillerthal, owned by Prince Fürstenberg, are estimated to shelter from 2,500 to 3,000 head of chamois.

Besides these private preserves there are innumerable parochial preserves belonging to villages and hamlets, each house-owner having the right to shoot over a district of vast proportions.

The villages of Brandenberg and Steinberg, in North Tyrol have, for instance, the shooting over not less than 48,000 Joch (about 80,000 acres) of the very best shooting ground to be met with in Europe, excepting perhaps some of the Scotch preserves, that cost their owners thousands of pounds, while here the concern pays each of the co-owners according to his annual bag.

For the benefit of those of my readers who have never seen a chamois I may give the following abridged description of the animal.

Somewhat larger than a roedeer, a chamois weighs when full grown from forty to seventy pounds. Its colour, in summer of a dusky yellowish brown, changes in autumn to a much darker hue, while in winter it is all but black.

The hair on the forehead and that which overhangs the hoofs remain tawny brown throughout the year, while the hair growing along the backbone is in winter dark brown and of prodigious length ; it furnishes the much-prized 'Gamsbart,' literally 'beard of the chamois,' with tufts of which the hunters love to adorn their hats.

The build of the animal exhibits in its construction a wonderful blending of strength and agility. The power of its muscles is rivalled by the extraordinary faculty of balancing the body, of instantly finding, as it were, the centre of gravity. A jump of 20 or even 25 feet down a sheer precipice on to a small pinnacle of rock, the point of which is smaller than the palm of a man's hand, is a fact of constant recurrence in the course of a chamois' flight.

With its four hoofs, shaped like those of a sheep, but longer and more pointed, and of a much harder substance, converging together, it will occupy this position for hours, watching any particular object that has attracted its notice.

The marvellously keen sight and scent of this fleetest of the antelope species is equally a matter of wonder. A chamois, frightened by some un-

K

usual sound or sight, and dashing down the precipitous slopes of the most inaccessible mountains, will suddenly stop, as if struck by lightning, some yards from the spot where recent human footprints are visible in the snow, or when, by a sudden veering of the wind, its keen scent has warned it of the vicinity of a human being.

It is obvious that the chase of an animal gifted with such extraordinary powers of locomotion, endurance, and with an amazingly keen scent, detecting danger at a great distance, requires corresponding faculties on the part of the hunter.

The power of undergoing great fatigue, privations, and cold, a steady hand, and a cool clear head and nerves, are the 'sine quâ non' that go to produce a chamois stalker ; and it is just the knowledge and consciousness of possessing these qualities that in nine cases out of ten furnish the mainspring of the hunter's passion.

The hunter must rely entirely upon himself. Neither man nor dog can be of service to him, and no fear of hunger, cold, and the yawning abyss at his side should make him waver or turn.

When following his game high up in the grand solitude of the sublime giant peaks, he is lost to man and the pursuits and passions that sway other men's destinies. He is entirely carried away by the excitement of the sport ; he crosses fields of snow without thinking of the chasms which are hidden under that treacherous cover ; he plunges into the most inaccessible recesses of the mountains ; and he climbs and jumps from crag to crag, and creeps along narrow bands of rock overhanging terrible precipices, without once thinking how he can return. Night finds him high up, seven or eight thousand feet, perhaps, over the tiny little valley that contains his poor dwelling. Alone, without fire, without light, without any sort of shelter, he has to pass the cold night close to glaciers and vast snowfields.

The chief characteristics of a chamois hunter's appearance might be comprised in the following short delineation : a gaunt and bony figure, brown and sinewy knees, scarred and scratched, hair shaggy, and hunger the expression of the face ; dark piercing eyes, marked eyebrows, a bent eagle nose, and high fleshless cheek-bones.

The shirt open in front displays the breadth of the hairy mahogany-hued chest, while the strong and bony but fleshless hands, with talon-like fingers constantly bent, clutch the long and stout alpenstock.

The chamois and its chase has for ever been a rich mine of anecdote and myth. The elder Pliny,

the great Roman naturalist, gives us in his Natural
History a striking proof of the gross superstition
which attached to this animal in old times. Among
other distinctive peculiarities with which he invests
the chamois he declares that the blood of the
chamois possesses great healing powers for several
diseases, such as consumption and low fever ; but
for one ailment in particular its qualities are a
specific, namely, ' the loss of one's intestines,' as he
terms a malady which we must hope, for humanity's
sake, has since disappeared from the long list of
mortal sufferings. He closes his remarkable description of the animal with the somewhat mysterious disclosure, that the blood of the buck used
in a certain manner softens the diamond into a
sort of kneadable paste. ' This latter piece of important information,' the author adds, ' has recently
been doubted by sceptics.'

One cannot but be amazed that such absurd-
ities were devoutly believed for many centuries ;
but it must be a source of even greater wonder to
read in modern descriptions of the chamois whole
pages of nonsense not a whit less astonishing.
One recent author, for instance, maintains that the
hunter rarely shoots, but drives his game into
places from which further retreat is impossible ;

he then draws his knife and 'puts it to the side of the chamois, and the animal of its own accord pushes it into its body!'

The recently-invented trick of 'intelligent' hotel-keepers in Switzerland, of placing a stuffed chamois on some crag a couple of hundred feet over the hotel, and then pointing it out to unsuspicious tourists, cannot throw much light on the chamois's habitat, however pleasant it must be to sightseeing cockneys to be able to eat their 'Gamsbraten' and drink their pint of sour Swiss wine under the very nose of a royal chamois buck.

No doubt such a make-believe sight tends to confirm the innocent tourist in his conviction that he is in the midst of the glorious snow and glacier-covered Alpine peaks, watching the sportive chamois; and we well may suppose that the prospect of astounding willing ears on · his return home with narratives of the numerous herds of chamois he has closely watched, gladdens his heart.

Returning to Tyrol, where such devices are as yet unknown, and I hope will remain so for many years to come, we must glance once more at the chamois stalker.

His motives, even if he is a poacher, are not mercenary. It is the chase itself which attracts

him, and not the value of the prey ; it is the excitement and the very dangers themselves which
render the chamois hunter indifferent to most
other pursuits and pleasures. The glorious Alps,
the grand stern solitude reigning around him, the
gaunt peaks, and not least the exhilarating influence of the clear, bracing air, that renders motion
and exertion a pleasure, instil in him an inordinate
love for the solitary sport. 'A chamois stalker
who would exchange his life for that of a king is
not a genuine chamois hunter,' I have been told,
not by one, but by twenty ' Gamsjäger ; ' and were
I to call my own feelings into question I must
corroborate this sentiment.

Before giving my readers any instances of my
own experience of the kingly sport, I must notice
an interesting instance where a woman, urged by
love, shared the perils and hardships undergone
by her lover, a noted poacher, and exhibited a
remarkable spirit of ₁fortitude under the most
trying circumstances.

Those of my readers who have ever visited the
interesting old castle ' Tratzberg,' near Jenbach,
on the Kufstein-Innsbruck line of rail, will no
doubt have been struck by the very remarkable
workmanship of divers groups of game in life-

size, carved in wood, that ornament the hall and passages of the castle.

They display to the eye of a connoisseur great skill in their life-like imitation, and one is struck with the accuracy of every detail, be it the bend of a noble hart's neck or the graceful attitude of a roe-deer, or the exact colouring of the chamois' hair.

The man who, by dint of his rare skill, has thus pourtrayed game in their wild state, was once a noted poacher, and now has risen to be one of the best carvers in this part of the country.

The circumstances that brought about the transformation of a daring poacher—who, it is said, proved himself on more than one occasion a relentless foe of the keepers—into a skilful artist, are the subject of my brief biography.

Toni, for such is the Christian name of the ex-poacher, is a native of the village E——, in the Unter-Innthal, and the surrounding large and well-stocked preserves of a certain noble duke afforded him in his character of poacher the very best sport ; but, as a natural consequence, he ran the most deadly risk, every time he set out on his expeditions, of never returning home. A bullet, he well knew, was pretty sure to find its way into his body if he persisted in his reckless course.

Fortunately for him 'the course of true love' saved him from a violent death. Pretty Moidl, a daughter of a wealthy peasant in Toni's native village, had been for some time past the object of his fondest hopes and the subject of many a daring 'Schnaddahüpfler,' sung in the village inn on festive occasions.

Marriage between the poor penniless poacher and the daughter of the rich peasant was, of course, impossible, and so the two young people loved and sinned behind the backs of the parents.

In a short time the dire results of the free and easy love-making *à la Tyrol* began to show. The girl, terribly frightened by the thought of her parents' wrath, determined to elope with the choice of her heart.

When the white pall of snow had vanished from the adjacent peaks and mountains, and the balmy May sun was enticing the more venturesome peasants to drive their cattle to the verdant mountain slopes, Toni and his sweetheart suddenly disappeared one fine day from their village.

Nobody knew where they had gone, and the mystery grew darker when, some weeks afterwards, the report was spread that Toni had been shot in an affray with keepers.

It was not known where and by whom ; and

the keepers of course took good care to give evasive answers to any indiscreet questions on the subject of Toni's fate.

All this time our hero and his fair donna were inhabiting a disused woodcutter's hovel high up on the mountains, in a tiny and excessively wild mountain gorge, uninhabited save by the royal hart and agile roedeer.

For their sustenance they had to depend entirely upon the rifle of Toni : milk, bread, flour, or any other of life's most necessary commodities were beyond their reach.

One night, two or three days previous to Moidl's confinement, Toni failed to return from his daily raid in quest of game. The girl was in a sad plight. Too weak to regain the next inhabited valley, some eight or ten hours off, she was at her wit's end, and beginning to repent her bold step.

On the eve of the second day unfortunate Toni entered the hut. Bloodstained, hardly able to stand, and terribly weakened by the effects of a wound, he presented a sad spectacle to the loving eyes of his devoted girl. It seems that Toni had been tracked by the keepers, and while watching the approach of some roedeer, he received a ball right through the fleshy part of his shoulder.

Springing up, he was lucky enough to escape

his pursuers, and in his dread of having his retreat discovered he took the opposite direction, and thus foiled the suspicions of his antagonists.

Anxious to elude his foes, who he feared would institute a close search among the adjacent peaks and passes, he and Moidl left the miserable hut that very night.

A sort of cave, distant about two hours from their abode, was their goal. After a wearisome and perilous ascent in the dark night, they reached their new hiding-place just as dawn was breaking. Both had exerted their utmost strength ; he weak from loss of blood and the effects of his wound, she on the eve of her confinement.

The next day Toni set out in quest of game, and on his return towards evening with a chamois on his back, he found poor forsaken Moidl the mother of a babe. Being without means of lighting a fire, he could not even cook the meat, and for the first day Moidl had to find the necessary sustenance in the blood of the chamois, of which she drank about two pints.

The next morning Toni set out for a distant Alp-hut, where he hoped to find some matches and some cooking utensil or other. He was fortunate enough to find a boxful of the former and an iron pot.

The third day Moidl was already up and about, and with the aid of some water and the iron pot cooked some broth for Toni and herself.

The child born in such primitive and original quarters throve, and formed a fresh link between the two faithful lovers.

For eight weeks these poor creatures resided in the cave, and would have continued very probably till approaching winter obliged them to descend, had not an accident occurred to poor Toni.

On one of his raids he crossed the imaginary boundary line, running along a high ridge of mountains, which divides Tyrol from Bavaria. As he was returning, laden with a roebuck, two keepers from the Bavarian preserves and two keepers from the Tyrolese shooting grounds perceived him, and united their forces in order, if possible, to catch him alive. They succeeded only too well, and poor Toni was transported the following day to the next Bavarian town, some thirty or thirty-five miles off. There he was committed for trial; and the result was a sentence which condemned him to a comparatively long term of imprisonment.

Luckily for him he was brought to one of the model prisons near Munich, where he was taught the rudiments of drawing and carving; and when he left the penitentiary he had imbibed a strong taste for carving from nature. After several years' imprisonment he returned home and set up a primitive sort of workshop.

Moidl, on the contrary, finding that Toni did not return from his shooting expedition, waited for a few days longer, and then descended to civilised valleys. Afraid to return home with the proof of her guilt in her arms, she turned her back on Tyrol and went on foot to Tegernsee, a lake in Bavaria, a good distance off. There she found kind people to take care of her child, and to her great joy she learnt too that her Toni was not shot, but only imprisoned. After stopping a few months with her child, she returned to her native village and re-entered her paternal home as if nothing extraordinary had occurred. None of her family, and none of the natives of the village, ever learnt the details of her exploit, and very probably never will.

To return to Toni's career. The owner of Castle Tratzberg, Count E——, happened to see one of the heads of a chamois turned out by Toni, and perceiving therein the undoubted traces of great skill, sent him, at his own expense, to a celebrated Bavarian school for carving in wood from nature. Here Toni stayed a considerable period, and left it the finished artist he now is.

Now to instances of my own experience of the noble sport of chamois stalking.

Delightful old Schwaz, a quaint village dating its existence back to the early Middle Ages, situated on the right-hand bank of the swift Inn, has been for years a favourite starting-point for my chamois-stalking expeditions.

Right opposite the quaint old-fashioned houses forming the main street, and on the opposite side of the valley, the high and terribly steep ' Vompergebirg ' rises in one unbroken mass up to nearly 9,000 feet over the level of the sea.

Far in among the oddly-shaped pinnacles which rise to even a greater height than the front peaks, which are partly visible from the Inn valley itself, there is a deep and narrow glen, and snugly ensconced in it is a small log-hut, surrounded by a lovely grove of beech-trees. Built for the convenience of the gamekeepers of the vast surrounding preserves, who have to be constantly on the watch lest poachers, reckless of the terrible risk they' run, should enter them, it has been many scores of times my night-quarters.

It was towards the end of October, 187—, that a six hours' walk from Schwaz brought me to the Zwerchbachhütte, the name of the hut I have just described. My kit for chamois-stalking expeditions is of a somewhat bulky nature, and generally a weight not far short of eighteen pounds has accumulated by the time a big piece of bacon, a dozen or so of hard-boiled eggs, bread, tea, and sugar, a flask of Kirschwasser, a telescope, and that most important of culinary implements, a small iron pan with a hinged handle, have been packed into my 'Rücksack,'[1] the weight lying to a great extent against the small of the back.

Having left Schwaz at daybreak, I had reached the hut and cooked my simple repast by half-past ten o'clock. I had thus ample time for an afternoon stalk. Leaving everything save my rifle, alpenstock, 'Steigeisen' (crampons), and telescope at the hut where I intended to pass that night, and even divesting myself of my heavy coat, so as to reach the heights of the mountains with as little loss of time as possible, I set out on my stalk.

As I looked up from the hut to the summit of the snowclad peaks, it seemed impossible that human foot could gain them; and yet, to have any chance with the chamois, I must be on the top of an immense crag some 2,000 feet above my head, in an hour, or at the latest an hour and a half.

---

[1] A sack of strong canvas with two broad leather straps, through which the arms are looped.

By a few minutes after three I had gained the aforesaid point. Night would fall at about six or half-past, and counting an hour to get down, I had still about two hours to spare.

Reconnoitring with my telescope the rising precipitous slopes of the adjacent peaks, I soon discovered a herd of nine chamois, amongst which I perceived a patriarchal buck.

As the wind came up from the valley—a matter of high importance, on account of the amazingly keen scent of the game—I had to decide to make a considerable round in order to weather them. After an hour's hard scramble I had gained the same altitude as that of the herd in view. Had the ground which now intervened between me and the game been a little less unfavourable, everything would have gone well; but the only means of getting within range of the wary animals was by creeping along a narrow ledge of about 2 to 2½ feet in width, that ran horizontally across the face of an immense wall of rock, at the other end of which the chamois were browsing on the stunted 'Latschen' that grew there.

The ledge was not more than 400 or 500 yards long, but I was obliged to proceed very slowly and carefully, for fear of betraying myself by knocking any of the small stones which littered the ledge

down the precipice—some two or three hundred
feet in height—which yawned at my side.

At last, after more than an hour and a half's

'AT LAST I GOT WITHIN RANGE.'

hard work, I managed to reach the end of the ledge,
and picking out my buck at about 160 yards, I fired.

Intently watching the effect of my shot, I saw

the chamois rise on his hind legs and fall over backwards, a sure sign that he was mortally wounded.

The charm and excitement which the successful hunter experiences in moments like this are not easily described. Certain it is that few other pleasures that life can offer are preferable to them.

Reloading my rifle, I hastened up to the spot, but found the buck had vanished. The colour of the blood which lay in a pool on the rock convinced me, however, that the game was hit hard, and could not be very far off.

Not till now, when it was too late, did the imprudence of proceeding so far by the waning daylight strike me. What should I do? Pursue the wounded buck, or try to return to the hut? A few moments' consideration showed me that long before I could reach the really dangerous places in the descent night would have fallen. In full daylight it required a very steady head and an extremely sure foot, as in most parts it was certain death to place one's foot an inch to the right or to the left of the jagged stones projecting from the rock, by the aid of which the ascent or descent could be accomplished. Thus I had to choose the more prudent course of patiently enduring the punishment of my rashness, which in this instance consisted in camping out.

L

Had I been provided with the necessaries for
so doing I should not have had any reason to dread
the approaching night ; but without a coat on my

MY NIGHT-QUARTERS AMONG THE ROCKS.

back, without blanket or anything to cover me,
and without a particle of food, the case was very
different, and I entertained some unpleasant notions
of the coming eleven or twelve hours.

Leaving the buck to his fate, I set about looking for a suitable nook or crevice which might offer some slight shelter. The waning daylight enabled me to find such a retreat in the shape of a small cave-like recess, which looked anything but inviting. The vast snowfields in close proximity, the icy-cold wind driving straight down from them, and an atmosphere considerably below freezing-point, did not add to my comfort. The only consolation left to me was my pipe, and before morning broke it had been filled and emptied many a time. At last the rosy tinge of the heavens, now unclouded by snow, which had begun to fall about midnight, assured me that my sufferings were coming to an end ; and never in my life do I remember greeting light with such feelings of gratitude as on that morning. My flannel shirt, saturated by perspiration the evening before, was frozen, and formed an icy coat of mail for my shivering body inside it.

Fortunately the snow lay very thin, so that it was easy to follow the gory tracks of the wounded buck. Half-an-hour's invigorating climb brought me to the place where the animal had evidently passed the night ; large pools of partly fresh and partly congealed blood marked the spot.

I had not proceeded more than a couple of hundred yards further up a narrow gorge when a shrill 'phew'—the chamois' whistle of alarm—brought my rifle to my shoulder, and levelled at the buck, standing on a crag projecting from the otherwise smooth surface of an immense precipice. The next instant my shot awoke the slumbering echoes of the ravine, and the buck came tumbling down the declivity, this time not to get up again.

On reaching the animal I found that my first ball had pierced its lungs. It seems hardly credible that an animal mortally wounded could continue its flight up the most dangerous passes and over chasm-parted crags, and that its steel muscles could carry it on and on after losing such quantities of blood. But so it is, a wonder to those who know the miraculous vitality and tenacity of life which characterises this magnificent little mountain antelope.

Brittling the game—that is removing the intestines and filling the cavity thus formed with twigs of a neighbouring 'Latschen' bush—I managed to fasten the buck, with the aid of my leather belt, to my back, and turned my steps homeward. I doubt very much if I could have reached the hut had I not had my trusty crampons on my feet.

The thin coat of snow covering the rocks made the descent of a doubly dangerous nature ; added to which I had a fifty-pound weight on my back, and naturally felt somewhat faint for want of

I PICKED MY STEPS BAREFOOTED.

food. In one place I was fairly compelled to divest myself of crampons, shoes, and socks, and pick my faltering steps barefooted over the projecting crags on the face of a perpendicular wall of rock, at the foot of which, some 2,000 feet below me, lay the hut, inviting one gigantic leap which would land me at its very threshold. At last, after one or two somewhat narrow escapes, I reached my asylum, and right glad I was that this descent, one of the most perilous I ever remember, had ended so satisfactorily.

By the time a hearty meal and a few hours' sleep on the soft and fragrant Alpine heather had restored my vigour, the afternoon had passed, and had it not been for a bright full moon, which promised to light me home, I should have remained that night in the hut.

Soon after sunset the full disc of the moon rose over a gap in the otherwise unbroken ridge flanking the gorge in which I was now walking homewards.

The huge gaunt forms of the peaks and crags, in many parts in deep mysterious shade, contrasted most charmingly with the glittering snowfields and ashy white peaks illuminated by the rays of a full moon. Now passing a cataract of white foaming water, glittering and gleaming as the moonbeams touched each distinct drop, then again traversing dense gloomy pine-forests, the tops of the trees tinged with silvery light, the rest dark and sombre ; now fording a turbulent rivulet, rushing down the declivity in headlong haste, then again crossing peaceful stretches of Alpine meadow-land dotted here and there with clumps of patriarchal pine-trees, my walk proved a delightful close to my expedition.

The reader, however, must not infer from this narrative that the lonely chamois stalker always meets with success at a cost of so little time and trouble as I experienced in this instance.

Droves of nine head of chamois are not to be met with in all parts of Tyrol, and often and often has it been my fate to be high up in the barren, terribly grand recesses of the Tyrolese Alps for days and hardly see a chamois ; or, at other times, an un-steady hand at the moment of firing has obliged me to traverse glaciers, snowfields, and passes to seek a distant glen or peak where the chamois had not been alarmed by the echoes of my shot.

Frequently two days elapse from the time of leaving the valley before a buck has been sighted and the line of attack resolved upon ; and then often, when after endless fatigue and danger the game has been nearly brought within range, the wind may suddenly veer, and a second later a shrill 'phew' of the alarmed chamois tells you that the fine scent of your prey has frustrated all your designs.

On one occasion, I remember, while hunting in the rugged 'Kaisergebirg,' I had approached a drove of six or seven chamois to within shooting distance, when the sight of a 'Steinadler' or golden eagle, which, circling right over my head, was allured probably by my motionless position *ventre à terre* for more than an hour, sent my game away in the twinkling of an eye and long before I had time to venture a long shot at the wary old buck who was keeping guard furthest off from me, and for whose approach I had been patiently waiting.

Another time, on the same mountains, I was imprisoned for two nights and one day on a pinnacle of rock by the accidental slipping of the rope which had enabled me to gain the eminence. The jump, or rather the drop, that eventually set me free was not much of a jump in any ordinary place, but here it was a very serious affair indeed. I had thrown the ill-fated rope, provided with a running noose, so as to catch any projecting particle of the rock, from a band of rock not more than twenty-eight or thirty inches broad, running horizontally across the face of a stupendous precipice four or five church steeples high. Now that the rope was gone I had to jump the height, up which I had hauled myself by means of the rope. The distance intervening between the band of rock and the point I was standing on was less than twelve feet in height, and deducting seven or eight feet which I could cover by lowering myself, and holding to the top by my hands, the actual drop, measured from the soles of my feet to the base of the miniature precipice where the narrow ledge projected, was about four or five feet. Nothing! if you have level ground to drop upon, and no yawning abyss at the side ; but here there were nine chances to ten that the drop would end badly.

It was only when the pangs of hunger on the morning of the second day, and the certainty of a lingering death by starvation, rendered me reckless of the terrible risk, and a sudden death seemed preferable to tortures slow and lingering, that at last I resolved to chance the drop.

Fate favoured me, and I alighted erect and firm on the narrow strip of rock that separated me from death. I had taken off my shoes and socks so as to prevent my slipping on reaching the ledge, at that part, if anything, shelving downwards. The slightest tremour of my knees, or the most minute giving way of my joints on alighting, would have resulted in the loss of my balance, and as there was nothing to afford me the slightest hold on the

smooth surface of the rock, I should have been
pitched head foremost down the abyss. My feet
were badly cut on the sharp stones on which I
alighted, and for weeks my little adventure was
recalled to my mind in an unpleasant manner;
I ought not, however, to complain of this insignifi-
· cant injury, considering I had a somewhat remarkable escape.

To show my reader that much time and exertion is expended and severe privations are vainly en-
dured by hunters while pursuing chamois in thinly-
stocked neighbourhoods, I may mention that in one
season I made the two expeditions I have just referred to, besides a third into the same range of
mountains, and in all these I did not fire one shot.

At other times, when the chamois are driven at
battues in the carefully-guarded preserves of either
of the three noble owners above-mentioned, a fairly
good rifle-shot, posted on an advantageous point,
can knock over from five to six chamois in the
course of a few hours.

In my humble opinion, and in that of every
sportsman who has once successfully 'stalked' a
chamois, the driving of chamois deprives the sport
of those highly attractive features, which act as an
ever-new, all-engrossing excitement on the mind of

the man who has once tasted its pleasures, beyond perhaps any other sport in the world.

It would seem to me that the wholesale slaughter of an animal that Nature herself has placed in the most sublime recesses of her creation, and endowed with such noble qualities and wonderful organisation, is a proceeding which a true sportsman ought not to countenance.

In the preceding pages I have endeavoured to give my readers an insight into the character of the chamois stalker, as well as to show the nature of the sport itself.

Manifold dangers and adventures of more or less peril, together with the hardships natural to the craft, are the fate of the chamois stalker, till perhaps some day or other he fails to return to his châlet, to his wife and to his little ones. A bullet from the rifle of a hostile keeper, or a treacherous bough, or a loose stone, or a false step pitches him to the foot of a precipice hundreds of feet in height, and years afterwards, perhaps, his bones are found, picked clean by the mighty eagle or by other wild animals of the Alps. A grand and silent grave, marked by a mighty tombstone set by his Creator himself, is only too often the last resting-place of a chamois stalker.

CHAPTER VI.

THE GOLDEN EAGLE AND ITS EYRIE.

NEXT to the poacher, the Golden Eagle (*Aquila Chrysaëlos*) and the Lämmergeier (*Gypactos barbatus*) are the two greatest enemies of the chamois and roedeer. Far less noble than the eagle in his proportions and build, the latter does not develope the exclusive appetite for blood and live flesh which distinguishes the eagle among the rapacious birds of prey.

The eagle, the tiger of his race, bears off his prey in triumph ; the geier very seldom attempts to remove it, but devours it on the spot—indeed his grasp is too feeble to permit him to manage effectually any but a comparatively trifling weight. The eagle, on the contrary, rarely touches carrion, and his terribly-powerful wings and talons enable him to carry off the strong-limbed chamois, or a full-grown goat, or sheep weighing considerably over thirty pounds.

If the animal singled out as his prey is too heavy, the eagle will swoop down upon it with resistless fury, and by mere force of the concussion will hurl it down the abyss at the brink of which it happened to graze or feed.

Several times have I had occasion to watch a golden eagle carrying off a young chamois or roe. The great weight of his prey would oblige him now and again to loosen his hold upon it while circling at a terrible height over ravine and peak. As it falls the eagle will dart after it, and catching it up in his claws, allow himself to sink for twenty or thirty feet by the mere impetuosity of his downward flight, and then, spreading his mighty wings to their widest, resume his circling ascent with his prey firmly clutched.

Tyrol, judged by what I have seen of it, does not harbour more than eight or ten pairs of golden eagles ; and Switzerland, I am told, is quite rid of these noble but terribly destructive birds of prey. The scale on which a pair of these birds will carry on their depredations among the game stocking the ravines and glens near the site of the eagle's home, the eyrie, is incredibly large. Quite a halo of celebrity is therefore thrown about the lucky shot who has brought down one of these royal highwaymen of the Alps. Far more exciting and difficult than shooting is the extraction of a young eagle from his nest or eyrie.

Eight or ten years ago I assisted in an attempt to rob an eagle's eyrie of its young inhabitants, in a remote glen in the Bavarian Highlands. Owing to the inadequacy of our means for approaching the goal, the attempt failed; but it left so vivid an impression on my mind that for four or five consecutive springs I was continually on the lookout for a repetition of this adventurous exploit. The difficulties of tracing one of the parent birds home to the eyrie are however so great, that the site of one of these royal homesteads is seldom discovered.

On my return to Tyrol from a tour in France and Spain in the first week of July 1872, the very first person greeting me at Kufstein, the frontier station, was destined to be the bearer of the most welcome news, that the site of a golden eagle's eyrie had been discovered in one of the side glens of the broad Inn valley.

Old Hansel, my informant, was one of the gamekeepers on a large imperial preserve close by

Kufstein. Some years previously I had on more than one occasion shared a hard couch with him under the stunted pines, when inopportune night overtook us high up in some Alpine wilderness, or near the glaciers and huge snowfields, while in hot pursuit of the chamois.

Hansel had heard of the discovery of the eyrie, and was just about to take train to the small railway station, about an hour's walk from the opening of the B —— valley, at the remotest extremity of which, some ten or twelve hours' walk off, the eyrie had been found.

Telegraphing to my friend, who was awaiting my arrival in Ampezzo in order to make some ascents in the Dolomites, that I should be detained for three or four days, I re-entered the train that was to carry us to our destination.

The next morning long before sunrise we were already on our eight hours' tramp to our goal for that day—the small cottage of a drift-keeper, in close proximity to the very wild and well-nigh inaccessible ravine which was to be the scene of the coming adventure.

Few of my fellow-travellers of the day before would have recognised me as the town-clad through passenger from Paris to Kufstein. An old time-

worn country-made shooting coat of the very
roughest frieze, short leather trousers, as patched
and discoloured as the poorest woodcutter's, gray
stockings, displaying to the critical glances of the
natives my knees still bronzed from the exposure
attendant on a long course of Alpine climbing in
the previous years, and a seasoned hat, which had
been originally green, then brown, and had now
turned gray, on my head, would, I presume, at least
have rendered recognition a matter of difficulty.

Tomerl, the keeper of the wood-drift, was an
old acquaintance of mine, whose qualities as a keen
sportsman had shone forth when, four or five years
previous to the date of the present exploit, I had
quartered myself for a month in his secluded habitation, spending the day, and not infrequently also
the night, on the peaks and passes surrounding his
modest cottage.

To buxom Moidl, his pretty young wife, I
was also no stranger, and her smile and blush on
welcoming us, assured me that she still remembered the time when, reigning supreme over her
father's cattle on a neighbouring Alp, she had
ministered on more than one occasion to the wants
of the young sportsman who sought a night's shelter in her lonesome châlet (distant at least five

hours' walk from the next human habitation), in which she, a young girl of nineteen or twenty, did not shrink from playing the hermit for four or five months of the year.

Many a merry evening had I spent in the low oak-panelled 'general room' of Tomerl's cottage when he was still a gay, though middle-aged bachelor. No changes had since been made in the aspect of the apartment.

In one corner stood the huge pile of pottery, which, being used for heating the room, one might by mistake have termed a stove. Over this singular masterpiece of pottership, about two feet from the ceiling, was fixed a sort of shelf, four feet broad and six long. This represented the nuptial couch of the couple. ' In winter,' as Tomerl laughingly remarked, ' it is warm and cosy, and in summer it has the advantage of being a bed taking up but little space.' Running the whole length of two walls of the room was a broad bench, in front of which were placed the two strong oak tables, round which on Sunday evening such of the woodcutters as were at work in the near neighbourhood used to congregate, to laugh, sing, and quarrel over the glasses of home-brewed ' schnapps ' which Tomerl,

in utter defiance of the excise officers, ventured
to sell to them.

We arrived at Tomerl's cottage just as they
were beginning their twelve o'clock dinner. A
second edition of a huge iron pan filled with the
savoury, but somewhat too greasy Schmarn very
soon made its welcome appearance. Amid laughter
and merriment our repast came to an end, and
we began our confab as to the best means of attain-
ing our end, viz., the young eagle.

Two woodcutters, whom we had found seated at
one of the tables on our arrival, were despatched
to a neighbouring woodcutter's hut in order to
fetch the four inhabitants of the same, whose
presence at our consultation was a matter of vital
importance.

As it was Saturday, they had knocked off work
in the course of the afternoon, and had adjourned
to the hay-loft for a few hours' sleep, prior to set-
ting out for a poaching raid to the distant Bavarian
preserves.

On learning the object of my presence they
immediately hurried down to Tomerl's cottage, and
half an hour later I was in possession of all the
facts and information regarding the whereabouts
of the 'horst,' or eyrie, the difficulties which would

have to be surmounted, and the manner in which the discovery had been made.

Their vocation as woodcutters, it seems, had brought them, while decimating a forest distant about nine miles from the hut, to the extreme end of a narrow and wild mountain ravine, just opposite the eyrie, which, with the usual parental care, was built in one of the small crevices by which the Falknerwand, a peak the side of which towards the valley is a perpendicular wall some 900 or 1,000 feet in height, is riven.

The evening was spent in discussing the details of the exploit, and getting our various implements in order.

We were up in the morning by three, and an hour later we were ready to start.

Our force consisted of six woodcutters—who were only too glad to give up their poaching expedition for the more exciting one on which we were now bent—Tomerl, Hansel, and myself. After shouting a last 'joddler' to his wife, who returned the greeting with her clear, bell-like voice, though her heart was doubtless beating fast under her smartly-laced bodice as she waved us a last adieu, Tomerl took the lead of our long file.

Three hours later we had reached the base of

the wall, the site of the eyrie. I immediately saw
that, besides being a more adventurous affair than
I had anticipated, nothing could be done from this
side of the peak. Indeed the precipice seemed
not only perpendicular, but actually inclining
forward in its upper part, and this impression
seemed to be borne out by the fact of our finding
close to the base numerous blackened remains of
fires which had been lit under the shelter of the
cliffs by belated keepers, or, what seemed even
more probable, by poachers.

By a circuit of considerable length we finally
gained the summit of the peak, and throwing down
our various burdens, we began to reconnoitre the
terrain, which we did *ventre à terre*, bending over
the cliff as far as we dared.

Great was our dismay on perceiving, some
eighty or ninety feet below us, that a narrow rocky
ledge, which had escaped our notice when looking
up from the foot of the cliff, projected shelf-like from
the face of the precipice, and shut out all view of
the crevice which we supposed contained the
eyrie.

After consulting some time we decided to lower
ourselves down to this rock band, and make it the
base of our further movements, instead of operat

ing, as we had intended, from the crest of the cliff, where everything, but for this obstacle, would have been tenfold easier. Posting one of the men at the top of the crag to lower our heavy fifty-fathom half-inch rope by a cord, after we had gained the ledge, we descended one by one, hand over hand, to the site of the coming exploit.

The ledge was of varying breadth ; in some places it was less than two feet, in others again it widened to about seven or eight feet; but at the place right over the crevice, where the men handling the rope had to take up their position, it was from three to four feet in width. Of course this was a somewhat embarrassing circumstance, necessitating extreme caution in all our movements, besides having the disagreeable feeling of standing at the very edge of a yawning gulf some eight hundred feet in depth, and nothing to lay hold of for support but the smooth face of the rock.

We had lowered ourselves in the order which the men had to occupy during the ensuing operations. First came Hansel, then the five remaining woodcutters, then myself, and finally Tomerl, the first and the last provided with their rifles.

On reaching the ledge we immediately began operations by driving a strong iron hook into the

solid rock at a point some two or three feet above
the ledge.   Through this hook the rope was
passed, one end pendent over the cliff, and to ob-
viate the peril of its being frayed and speedily
severed by the sharp outer edge of our platform,
we rigged up a block of wood with some
iron stays, to serve as an immovable pulley.
By means of the hook the rope was directed
sideways to the spot where the men told off for pulling were standing in single file, a space of about
three feet between each.

After completing our arrangements I turned
my attention to the broad leather belt, similar to
the one worn by our fire-brigade men, that was to
fasten me to the rope.

To fasten the belt round my waist, to run
the rope through the strong iron ring in front of it,
and knot it securely to a strong piece of wood,
my seat, were our next proceedings.   This manner of fastening oneself to a rope is preferable
to the orthodox way of binding waist and both legs
to the rope, as it impedes free movement far less ;
and even if I were to slip off my wooden horse I
could not fall, the wood preventing the rope from
passing through the ring.

A large hunting-knife was in my belt, a small

but powerful Smith revolver in my pocket, and in my hand a long pole, shod with iron at one end, and at the other fitted with a strong boathook, which we had forged the night before in the miniature smithy of Tomerl's cottage.

The five woodcutters took hold of the rope, while the two keepers, *ventre à terre*, began their duties as my guardian angels by cocking their trusty rifles, in case of any attack of the old eagles while I was engaged in my work of spoliation. On their watchfulness and on their unerring aim my life would, in case of such an emergency, depend, just as much as on the muscular arms of the five shaggy-headed woodcutters.

Laying hold of the pole, I gave myself a gentle push, which sent me clear of the edge into space. Although it was not the first time I had been in a similar position, the prodigious height was, for the first two or three minutes, not without a sort of exciting effect on my nerves.

Five minutes later I had quite recovered, and enjoyed the novel position of hanging on a rope, scarcely thicker than a man's finger, over an abyss of nearly 1,000 feet in depth, quite as much as any new and hitherto unknown sense of danger charms the minds of men fond of rough Alpine climbing and mountaineering in the strict sense of the word.

The descent lasted not more than ten or fifteen minutes, and when I arrived opposite the crevice, where the existence of the eyrie was plainly indicated by a mass of dry sticks and refuse of all kinds strewn about, I stopped further progress by two distinct jerks at the signal-line.

The distance separating me from the eyrie was, owing to the projecting nature of the ledge on which the men holding me were standing, and to the overhanging formation of the entire precipice, some ten or twelve feet; but by the use of my pole, the hook of which I caught on a projecting stone, this difficulty was soon overcome.

At first the bulwark of dry sticks, the interstices between them being filled with dry moss, prevented my seeing anything. Cautiously crawling up an inclined slab of rock that led to the eyrie, and slowly raising my head over the side of the latter, while with my right hand I guarded my head and face against any attempt of the young eagle to attack me, I looked in. My surprise and pleasure on finding not one, but two young eagles therein may be imagined.

A peal of shrill shrieks, and sundry rather omi-

nous-sounding hisses greeted my unlooked-for
appearance.

WHAT WAS MY JOY TO PERCEIVE TWO YOUNG EAGLES.

Vainly flapping their enormous wings, while
with their small but inexpressibly wild eyes they
kept staring at me, they opened their beaks—

hooked at the end, and already of an alarming
size and strength—to their widest extent, plainly
indicating that their breakfast hour was nigh.

Detaching the stout canvas bag with which I
had provided myself from my seat, I proceeded to
bag one of my young prisoners. While he was yet
struggling in the ample folds of the bag, which I
had thrown over his head, I pinioned his formidable
talons, and then, unbagging him, I proceeded to
secure his wings and beak by means of a piece of
cord. I then deposited him in the bag, which, although a good-sized one, he entirely filled out, thus
excluding the idea of putting the other bird into
the same receptacle. As it is a rare occurrence
that two young eagles are found in one eyrie, I was
unprovided with a second bag, and consequently
was placed in a fix regarding the means of securing my second prisoner. After a good many
ineffectual trials I at last managed to secure him by
flinging my coat over him and then slipping a running noose over his feet, after which it was easy
enough to bind and prevent him from doing any
mischief.

The bag containing the first bird I tied to the
signal cord hanging by my side ; the other I resolved to carry up in my hand, there being little

danger of his hurting me, if the cords of his
shackles held out against his vigorous efforts to
get free.

I was glad to get out of the eyrie after having
brought my expedition to this successful termination, for the stench created by the putrifying
flesh strewn by the parent birds about the adjacent
rocks was something dreadful and overpowering to
any senses more delicate than those of a bird of prey.
These relics, which I had the curiosity to count, consisted of a half-devoured carcase of a chamois, three
pairs of chamois horns, with corresponding bones of
the animals, the skeleton of a goat picked clean, the
remains of an Alpine hare, and the head and neck
of a fawn. Arranging myself on my seat, I fixed
the hook of my pole in its old place, and gave the
signal to hoist me up. The bird I held in my left
hand, while with my right I intended to let myself
gradually swing out till I reached the perpendicular
position.

As the sequel shows, I had reckoned without my
host. The first hard pull of the men at the rope,
nearly 200 feet over my head, which, contrary to
my instructions, was much too vigorous, wrenched
the pole out of my grasp, sending the latter to the
bottom of the precipice, and me, at a fearful pace

outwards. My position was, as anybody can imagine, most dangerous. The velocity of the retrograde movement would dash me with terrible force against the solid wall of the rock. There was only one way, and that a very dubious one, of saving myself. Fortunately my presence of mind did not forsake me in this critical moment, and I grasped at this only chance of preserving my life and limbs. Tilting the upper part of my body backward and my legs forward, I awaited the dreaded shock, taking of course the chance of my striking the rock feet foremost as the only way of saving myself.

The retrograde movement of the pendulum, to which my weight supplied the velocity, set in, and a second afterwards I was saved, having struck the rock with my feet, which, well protected as they were by my immensely heavy iron-shod shoes, were the only part of my body which could have effectually resisted the shock. The only bad result of the contact with the rock was a paralysed feeling in my legs, and a prickling sensation in my back and loins.

Need I say how thankful I was that I had not followed the promptings of my companions to take off, before leaving the ledge, my shoes and stock-

ings, in order to facilitate the climbing, which, as
we supposed, would be a matter of necessity to
enable me to reach the eyrie?

For what reason I refused to follow this advice, and do a thing which, in the course of my
chamois-stalking experience, I had done so very
often, is a mystery which I do not care to solve,
the fact of my life having been thus saved being
sufficient for me.

While the above incident occurred I had remarked that a dark object had flashed past me, so
close that I distinctly felt the pressure of the air,
and heard the whistling sound it created, as of falling from seemingly a great height. Thinking it
was a stone, I paid no further heed to it, my attention being moreover attracted to a sharpish gash in
my thigh, which the bird placed under my arm
had managed to inflict, although his beak was
bound with my pocket-handkerchief. Some loose
gunpowder strewn into the wound was an effectual if
somewhat painful cure, and it was only after having
applied it that I remarked that, instead of being
pulled upwards, I was quite stationary.

It appeared afterwards that the object which
flashed past me a few minutes before was the block
over which the rope ran, and which was of vital

importance in securing my safety. This of course I did not know at the time, and consequently my anxiety grew from minute to minute. An hour, and then another passed, and still I remained in my most helpless position.

The boulder of rock, projecting a few feet over my head, prevented any view of the ledge, and my shouts asking the cause of the delay received indistinct answers, the words 'patience' and 'wait' being the only intelligible ones.

These words might have been consoling, but for the fact that nature, to cool my impatience and make my position more ridiculous in her eyes, destined me for a cold bath, the water being supplied by one of those short but terribly grand thunderstorms which victimise Alpine regions in summertime.

My position exposed me to its full fury, without any possibility of escape, and ere long it burst over my head, drenching me to my skin in the first five minutes, while the lightning played about me in every direction, and terrific claps of thunder followed each other at intervals of scarcely a few seconds.

What heightened the danger, as well as the absurdity of my situation, was the chance that one or

both of the old eagles might return at any moment
under circumstances that must render a struggle, if
any ensued, a most unequal one. Supposing my
guardians to be still at their post, the distance of
the ledge was such as to make a shot at a flying
bird, large as it might be, anything but a sure one ;
and the tactics of the golden eagle when defending
its home do not allow of any second attempt.
A speck is seen on the horizon, and the next
moment the powerful bird is down with one fell
swoop. A flap with its strong wings, and the unhappy victim is stunned, and immediately ripped
open from his chest to his hip, while his skull is
cleft or fractured by a single blow of the tremen-
dous beak. Instances are however known in which
the cool, self-possessed ' pendant ' has shot or cut
down his foe at the very instant of the encounter.
Happily my own powers were not put to so severe
a test ; the old birds were that day far off, circling
probably in majestic swoops over some distant
valley or gorge.

I was forced, however, to be constantly on the
alert, and my impatience and perplexity may be
imagined as hours elapsed and there were still no
signs of my approaching deliverance. The storm
had long since passed over, and darkness was

settling down, when I felt a pull at the rope, and
my ascent, begun nearly four hours before, again
went on.

It was of the utmost importance that the whole
party should regain the top of the cliff before night
had fairly set in ; I therefore deferred on my arrival
at the ledge all questions till we had gained a place
of safety.   The heavy rope, fastened to the cord,
was hauled up by the man on the top, and after
it had been secured to a tree-stump, we swarmed
up without loss of time.

We had still before us a somewhat perilous
scramble in the darkness down the steep incline,
but the exhaustion attendant upon the fatigues and
privations we had undergone made it necessary
that we should first recruit our strength by means of
the food and bottle of schnapps we had brought
with us.   While we were doing justice to the bread
and bacon, and taking gulps of undiluted spirits,
the tale of the different mishaps of the day was
told, now by one, now by another of the sufferers.

It seems that as soon as the accident which sent
the block to the bottom of the Falknerwand was
perceived by the men engaged in hoisting me up
hand over hand, they desisted from their task,
lest the rope, now unprotected, should be injured

by the sharp-edged stones, and thus place my life in imminent danger. They communicated the mishap to the man on the top of the cliff, who immediately went to get a substitute. Descending to the base of the peak, he felled a young tree, and shaped a block similar to the one lost. As he was returning to the crest of the Falknerwand with the block on his shoulder, the thunderstorm overtook him, and one of the vivid flashes of lightning playing around him cleft and splintered a rock, weighing hundreds of tons, that had stood within thirty paces of him. He received no injury, except being thrown on the ground and partially stunned by the terrible concussion; but it was not till after a considerable time that he was able to rise and continue his ascent. What would have become of us, and me in particular, had the man been killed by the lightning it is difficult to say ; most probably starvation would have been our fate. The next human habitation, excepting old Tomerl's cottage, was eight or nine hours' walk from the Falknerwand ; and as Tomerl's wife did not know the direction of the eyrie, the chances of her finding us in time for mortal help were small, indeed so small that when I hinted the thought to my sturdy companions, the momentary gloom and dark frown on their shaggy brows told me but too plainly that they concurred in my dark anticipation.

Our meal ended, we placed our pinioned prisoners in a large hamper specially provided for their transport, and after some trouble contrived to manufacture two torches, in the ruddy glare of which we wended our steps down the steep incline to the bottom of the Falknerwand.

From some dry wood found beneath the sheltering precipice we made some more torches, and finally reached Tomerl's cottage at a late hour, rather worn and hungry but highly satisfied with our success.

A steaming 'Schmarn' and 'Speck' (bacon)— the latter a great treat for the men—soon appeased our hunger; the thirst, however, seemed to me to be of a more formidable nature, for it was close upon two o'clock when the last touch on the chords of the 'Zither,' which accompanied the final 'Schnaddahüpfler,'sent us up our ladder to the hayloft.

On my return next morning from my morning stalk, with a roebuck on my back, I had full leisure to look at the young eagles, who, released from their shackles, had been placed in a small barn, the door of which had been unhinged, and in its stead stout wooden laths fixed across the opening.

Before their fetters were untied, the wings had been measured, those of the hen bird being fully two or three inches larger than the wings of the cock bird, though the latter had the finer head. The hen bird measured 6 feet 11 inches in the span, and when full grown the breadth would very probably reach 8 feet, or 8 feet 6 inches.

The 'Aufbruch,' or entrails of my buck, together with two live rabbits, furnished a luxurious breakfast for the young captives. The rapidity with which it was despatched made old Tomerl, who was standing at my side watching the proceedings, shake his head, and ask me how on earth he could find the wherewithal to feed these two voracious babies.

A week after their capture they were 'feathered' for the first time. This process consists in pulling out the long, down-like plumes on the underside of the strong tail-feathers. These plumes, which, if taken from a full-grown eagle, frequently measure seven or eight inches in length, are highly prized by the Tyrolese peasants, but still more by the in-habitants of the neighbouring Bavarian Highlands, who do not hesitate to expend a month's wages in the purchase of two or three, with which to adorn their hats, or those of their sweethearts.

The value of a crop of plumes varies somewhat; generally, however, an eagle yields about forty florins' (4*l.*) worth of plumes per annum.

Six weeks after this incident I again found my way into the secluded B—— valley, and found that the hen bird had been sold to a neighbouring head-keeper of a large ducal preserve for forty-five florins (4*l.* 10*s.*). The cock bird I found alive and kicking. Being curious to see if his confinement had subdued his wild and ferocious spirit, I removed one of the laths, and entered the barn. An angry hiss, similar to that of a snake, warned me of danger, but too late to save my hands from severe scratches. With one bound and a flap of his gigantic wings he was on me, and had it not been for Tomerl, who was standing just behind me armed with a stout cudgel, I should have paid dearly for my visit.

I know of no instance in which human skill has subdued in the slightest degree the haughty spirit of the freeborn golden eagle. An untamable ferocity is the predominating characteristic of this noble bird, more than of any other animal. Circling majestically among the fleeting clouds, he reigns lord aramount over his vast domain, avoiding the sight and resenting the approach of man.

MY FIRST GLASS AT THE LITTLE INN.

CHAPTER VII.

AN ENCOUNTER WITH TYROLESE POACHERS.

A FOUR months' tour in quest of sport brought me, in the autumn of 1867, to L——, a small and entirely isolated Alpine village in the Bavarian Highlands, close to the Tyrolese frontier.

I do not know whether it was the result of a heavy day's work, wading, rod in hand, in the icy cold waters of the ' Isar,' or the knowledge that a certain fresh barrel of Munich beer was to be tapped—an event of no mean importance in the modest little inn of the village—which induced me, when night put a stop to my fishing, to seek a cosy retreat in the bar-room of the village Wirthshaus.

Hardly was I seated in my snug corner, right below the execrably daubed crucifix, adorning, as is the custom in the Tyrolese and Bavarian Highlands, the corner of every bar-room, when in

rushed, in an evident state of excitement, the
' Herr Oberförster '—head-forester of the surrounding royal game preserves.

My query as to the cause of his unusual emo-
tion was speedily answered.

One of his numerous under-keepers had at that
very moment brought him the news that four
' Wilddiebe,' or poachers, had been seen high up
on the mountains by two keepers, one of whom
had come down in hot haste to seek reinforcements
in order to capture the intruders.

Unquestionably, the head-keeper continued,
these poachers were the very same four Tyrolese
scoundrels who the year before had shot two of the
Bavarian keepers, and, hardly three months previously, severely wounded three others, who had
endeavoured to take them prisoners.

This was welcome news to my friend the Herr
Oberförster, who had on several occasions vowed
the destruction of that fearless and daring quartet
of Tyrolese, who in less than a year had killed or
maimed no less than five of his subordinates.

All the keepers who at that precise moment
were not out among the mountains were ordered
to assemble, and in a quarter of an hour six men,
eager to avenge their comrades' fate, were collected

in the head-keeper's cottage, whither I had accompanied him.

The evident fact that adventure of no ordinary
character would in all probability attend this exploit, naturally made me eager to witness the strife.
After some trouble I succeeded in persuading the
head-keeper to allow my accompanying the party,
of course, only as a mere looker-on.

To act as combatant on this occasion lay far
from my intentions, as, strange to say, my sym-
pathies were on the side of the Tyrolese, though, as
I have related, a twofold manslaughter was laid to
their door.

The deadly feud and animosity existing between the Tyrolese and Bavarian Highlanders since
the time of the French wars in the beginning of
the present century has by no means died out, but
flares up on frequent occasions.

The Bavarian preserves, well stocked with
game, but rigorously guarded by small corps of
game-keepers, aided by the rural policemen or
gendarmes, are looked upon by the Tyrolese living
close to the frontier as their legitimate sporting
ground ; and it is just on these occasions, when hostile parties meet, that the deadly animosity of the

Tyrolese poacher to the Bavarian keeper, and *vice versâ*, leads to murder and manslaughter.

To these two circumstances, and to the fact that the Tyrolese, inhabiting mountain recesses, have an innate love of wild sport, we must attribute the frequent encounters, resulting in the death either of the keeper or the poacher.

They are by no means moved to this dangerous game by any motive of gain, but simply by that love of free nature and the excitement of the perilous chase, which He who created the chamois and He who piled the mountains and glaciers upon each other has placed in their hearts, like the apple-tree in the Garden of Eden.

Thus it frequently happens that a young fellow, not content with the sport which his own mountains afford, leaves his home, an isolated châlet on the Tyrolese-Bavarian frontier, crosses the mountains, and, entering the forbidden land, fails one day to return to his home. A deadly shot from behind some ambush, a cry of anguish, and the poor fellow has paid the penalty of death for a crime which, even were it to come before a court of justice, would be punished with but six or nine months' imprisonment.

The body of the unhappy poacher, if it has not

A POACHER'S DEATH.

fallen down the yawning abyss, at the side of which he was walking, unconscious of danger, is pushed down into its deep and silent grave by the ruthless hand of the slayer, the gamekeeper, who, not caring to risk life and limb in a struggle with his foe, removes him from the face of God's earth by a cowardly shot.

Of late years this feeling of mortal enmity has somewhat abated ; but at the time I am speaking of, some seven or eight years ago, inquiries respecting the mysterious disappearance of a young Tyrolese from his native village or solitary châlet-home were invariably met by a shrug of the shoulders and ' Shot by the Bavarians.'

But to return to my narrative.

Our party, consisting of the head-keeper and six of his men and myself, were, after making some necessary preparations, ready to start.

With some bread, bacon, and a flask of 'Kirschwasser' in my bag, and with my revolver, in case of emergency, in my pocket, I joined the rest, who had already left the head-keeper's habitation.

The man who had brought the alarm led the way, then followed the Oberförster and his other men, and I brought up the rear.

The night being pitch dark, and our way lying

up some very awkward ledges and along some deep precipices, our progress was naturally slow, and the rain, which soon after our departure came on, did not serve to raise our spirits. Walking, and in many places creeping along on our hands and knees, we spent the best part of that night before we reached the spot where the two keepers had parted, one to give the alarm, the other to continue his watch on the movements of the poachers.

We were astonished to find no one there, and our undertone calls for ' Johann '—the keeper— remained unanswered.

All of a sudden our whispered consultation was interrupted by a low stifled groan, uttered apparently by a human being close by.

Fearing that this was part of a subtle stratagem of the poachers, who, we were now convinced, had discovered Johann, and intended by their groans to entice us to approach their ambush, we remained quite quiet for the next hour, till day began to break.

What dawn disclosed to our eyes, the reader will be astonished to learn.

Not thirty paces from the spot where we lay was poor Johann, divested of his coat and securely

JOHANN BOUND TO A TREE BY POACHERS.

pinioned to a pine-tree. With his mouth gagged, his face besmeared with blood, his rifle, broken at the stock, at his feet, he presented a sorry spectacle.

To cut him loose, force some spirits down his throat, and bind up his bleeding wounds was the work of a few minutes. When sufficiently recovered to speak he told us that while he was at his post, his gun had slipped from his hand, and, striking a rock, the charge had exploded.

The poachers, then not more than 400 yards off, just across a narrow but deep gully, at first imagined the shot was intended for them; but seeing nobody, they cautiously approached, rifle in hand, the spot where poor Johann had hid himself under some brushwood, afraid to move.

Searching the place, they soon discovered him, and threatening him with immediate death, they pinioned the poor fellow to the next tree.

His life hung upon a thread during the next five minutes, while the Tyrolese were deciding the fate of their prisoner.

The defenceless man must have moved their pity, for they took their departure soon afterwards, after inflicting with their iron-shod Alpenstöcke some painful prods on their hapless victim.

Had their prisoner been one of those keepers whom they suspected of picking off any of their comrades, a murder would have undoubtedly preceded their departure.

Watching his foes' movements as long as the waning daylight had allowed, he was convinced, by the direction the four men had taken, that they were encamped for the night in an Alp-hut not more than half-an-hour's climb distant, wholly unconscious of the fact that they had been seen by a second man, who had in a comparatively short time brought overwhelming odds against them.

As it was the month of October, and the Alp-hut, situated high up on the mountain, was occupied only during the three summer months, we were convinced that the hut was untenanted, thus affording a welcome night's shelter to the poachers.

It was now, naturally, a matter of the greatest importance to surprise the men while yet in the hut, and though, as Johann informed us, three of them had each a chamois on his back, they would not in all probability leave the hut for their return homeward before seven or eight o'clock.

Giving the necessary instructions to his seven

men—Johann was sufficiently recovered to join the party—the Oberförster and his little army made for the hut as fast as they could, while I was to gain, by a somewhat circuitous route, a little eminence right over the hut, whence I might overlook the whole scene of the coming combat without incurring any risk.

Half-an-hour's scramble brought me to the height, and on looking down the wreath of smoke curling up from the opening in the roof of the hut intimated that the poachers were still within, probably cooking their breakfast before starting on their perilous return over the frontier—in this instance an imaginary line running along the heights of the snow-covered ridge of mountains, rising in one sublime wall from the plateau on which the Alp-hut stood.

My post enabled me to see every movement of the eight men as they cautiously approached the hut, hardly 400 yards below me.

When about 150 yards from the châlet they divided, it being the intention of their leader to station one man at each corner of the hut while the remaining four keepers were to advance to the closed door.

They had hardly walked a few paces, when a

thundering ' Halt ! or we shoot,' from the poachers
within the hut brought the advancing force to a
sudden standstill, and, throwing themselves flat
down, they instinctively sought shelter behind some
trees and rocks which were lying around.

Caged, undoubtedly the poachers were, but by
no means caught.

To dislodge four resolute, well-armed men, dead
shots, from a bullet-proof log-hut standing in the
centre of a flat piece of ground, is by no means
an easy undertaking. The Oberförster, convinced
against his will of the impossibility of bringing
about a favourable result by force, unaided by
subtle stratagem, withdrew his men to a safer place,
whence the hut could be watched, without being in
imminent danger from the enemy's rifles.

At the trial of the poachers, who subsequently
were made prisoners, it appeared that the silent
man, attired in the garb of a cowherd, who was
sitting in the dark corner of the bar-room the previous evening, while the Oberförster related the
news of the poachers having been seen, had acted
as informant.

This man turned out to be a native of the next
Tyrolese village, and, without being in the least
connected with the poachers, he had, from mere

spite to the hated Bavarians, warned his country-
men of the approaching surprise ; too late, however,
to enable them or him to escape to their own side
of the adjacent peaks.

This of course explained the whole thing.   As
I was convinced that the head-keeper would post-
pone until night all attempts on the hut, I decided
to leave my post, and by a roundabout route join
the small but valiant army encamped barely 600
yards from the object of their continued watching.

On reaching them I found that one of the keepers
had been dispatched back to L——, and on my
inquiring the reason of such an arrangement, at a
time when every man was needed, I was informed
by the leader that he intended to take the hut by
assault at nightfall, and for this purpose needed a
bag of gunpowder to remove the barricaded door,
and thus enable the assailants to gain the hut
with comparatively little danger.

A very easy job it may seem to take by assault,
with a force of eight men, a simple log-hut, defended by just half that number; but when you come
to consider the substantial manner in which these
châlets are built, the immense thick door, iron-bound
and fastened by a huge beam drawn across it from

the inside, and the resolute, dare-devil character of the defenders, the reader will understand the difficulties with which the assaulting force had to cope.

Soon after sunset the keeper returned, accompanied by a confrère, whom he had found at home.

Soon afterwards, when it was sufficiently dark, we completed our arrangements.

The dangerous task of placing the gunpowder bag near the door of the hut, devolved on a volunteer, a keeper whose brother had been shot by Tyrolese poachers some years before.

Slowly creeping along, the man gained the door in safety, and placing the bag against the latter, lighted the slip of tinder which was to ignite the charge, consisting of four pounds of gunpowder.

A second later two shots from the hut made us tremble for the life of the brave volunteer.

All of a sudden a huge bright flame shot up, illuminating with a vivid light all surrounding objects. A terrible explosion followed, and a second later the eight men had, with one impetuous rush, gained the hut, and were pouring in through the breach produced by the explosion.

A shot, a second one, followed by a third discharge, intimated that the struggle inside that narrow log-hut was waging fierce and hot.

At this moment a dark object rushed past me up the incline on which I was standing.

A bullet, whistling past me in unpleasant proximity, induced me to throw myself down, while two of the keepers, in hot pursuit of the decamping poacher, nearly stumbled over my prostrate form. Another shot, and the hot and fierce fight was over.

On entering the hut by the doorway, now a large and ill-shaped breach in the timber, my attention was first attracted by the Oberförster stooping over the body of a man lying full length in the centre of the hut. The uncertain light of the fire in the open fire-place prevented my recognising the body till quite close to it.

It was old ' Berchtold,' one of the most trusty subordinates of the head-keeper, shot through the body. The poor fellow was apparently in a dying state.

Two of the other men were in the act of placing the gigantic form of a poacher on the table, while the remaining keepers were either busy binding up a wound in the arm one of their comrades had

o

received, or pinioning the only other poacher then visible.

But where were the remaining two keepers and the two poachers, who, as we supposed, had been sheltered in the hut, in addition to the two now before us? And who was that miserable object sitting, or rather crouching in the corner of the fireplace, with his hands in his lap, staring sullenly into the fire? These were all questions which arose in my mind while I was busying myself with the wound of the poacher stretched out on the table.

Before I was able to inquire the two missing keepers returned, holding between them a third 'Wilddieb,' whose face, originally blackened with soot, to disguise himself, was now, by the action of the blood trickling from a wound on the forehead, restored, in many parts at least, to its original colour.

Through all this excitement we had entirely forgotten the brave fellow who had fired the gunpowder, which had done such good service in clearing the way for the assaulting force.

On my reminding the Oberförster of their negligence a search was ordered, and the man was ultimately found, not twenty paces from the hut, in an insensible condition.

On examining him we found that a ball had grazed his head, and although it had rendered him insensible, he was not much hurt.

When the several cases had been properly attended to the question arose, What had better be done with those who were more seriously injured?

This point was not soon, nor easily decided. Old Berchtold was without doubt of all the wounded the one requiring most the aid of a doctor. The poacher on the table was sinking rapidly; but the two keepers, one wounded in the head, the other shot through the shoulder, and the poacher taken prisoner while attempting to escape, although not very seriously injured, would all be better for a more scientific dressing of their wounds than we were able to bestow on them.

It was decided, therefore, to start homewards as soon as a serviceable litter, for the transport of Berchtold, could be put together.

The rest of the wounded, and the poacher who had come out of the fight without a scratch, were to accompany the litter, while the dying poacher was to be left behind, his end being an affair of a few hours at the most. One of the keepers was to remain behind to watch over him, as well as over the mysterious man who had been found in the hut, and whom the Oberförster determined to detain till the arrival of the Government commission, which was to investigate the whole affair.

Two six-foot long Alpenstöcke, with a blanket and some branches of a pine-tree, furnished a capital litter. Passing a fresh bandage over Berchtold's wound, we placed him on it. Propped up with several coats, the poor fellow was better off than we could have hoped. Four keepers were told off to carry him, a task of considerable difficulty, owing to the steepness of the descent and the roughness of the path.

Next came the two injured keepers, followed by two poachers, both with tied hands ; the Oberförster, walking behind them, rifle in hand, vowing he would shoot the man attempting to escape, closed the file. One of the front carriers of the litter, and the keeper injured by the ball grazing his head, carried each a torch made of dry pieces of wood, between two and three feet in length, steeped in molten rosin.

While burning, these torches emit a brilliant and ruddy light, and as they are not easily extinguished by either wind or rain, they are preferable

to lanterns, which latter are rarely used in the
Tyrol or the Bavarian Highlands.

At the last moment I changed my mind, and
decided to remain in the hut for that night instead
of accompanying the 'train,' whose progress, tor-
turingly slow, on account of the wounded, would
in all likelihood only bring them to L—— towards
the morning.

On re-entering the châlet, after wishing the
departing file a safe journey, I found the poacher
in the same semi-conscious state in which I had
left him.

Lying there stretched to his full length, under
the glare of the pine-torch, stuck in between two
beams right over his head, he presented a most
painful spectacle.

His was a handsome, intelligent face ; his two
jet black eyes, fierce and angry in their expression,
when at intervals he opened them and bent a
piercing glance at the keeper, were the most remarkable features.

His hands, crossed over his huge, brawny chest,
clasped a rosary, which one of the keepers had
handed him, and the motion of his fingers, as now
and again they moved a bead, showed he was
praying.

Closely watching him from my seat at the fire-place, I perceived the pearly dew of death settling on his brow, and matting the locks of curly black hair which hung over his forehead. His gigantic frame, in which great power and agility seemed to be blended, appeared to stretch, while the muscles of his face began to twitch and distort his manly visage.

Presently he started up into a sitting posture, and in a high-pitched tone cried for his rifle. Stepping up to him, I offered to replace the bandage of his wound, which, loosely put on from the first, had been partially displaced by his violent movement. In a moment he fell back, apparently dead.

Both of us thought it was all over ; but I hardly had time to resume my seat, when all of a sudden he again started up, and with distorted face and shaking voice demanded a priest, 'for,' he continued, ' I cannot die till I have confessed.'

Hardly had he said these words when a stream of blood gushed from his mouth, and he fell back dead.

While yet speaking these words he had fixed his piercing eyes, unnaturally bright, with an expression of such deadly hate and mortal enmity

on the keeper, that when I looked round, when all
was over, I found the man with his hands before
his face, utterly stricken down by that one look of
unutterable animosity. It was only then that, by a
few words dropped by the man, I became aware of
the fact that he was the slayer of the poor fellow.

Though he had acted in accordance with the
letter of the law, empowering a keeper to shoot a
poacher who refuses to surrender, or endeavours
to defend himself, I have no doubt that dying
glance of his victim must have haunted him ever
after, warning him that he remained a mark for the
rifles of his victim's comrades, who would be only
too eager to avenge their clansman's death.

I left the keeper to his unpleasant meditations,
and returned to my seat at the fire.

All this time the mysterious man was crouching,
without even uttering a word, on the seat he had
occupied when first I entered the hut, some three
or four hours before. I addressed a few questions
to him ; but my queries remained unanswered, save
by a grunt and a sullen shake of his head.

Presently he rose, and going towards the doorway, was about to leave the châlet, when the keeper,
jumping up from his seat, restrained him, and told
him he was his prisoner. The man obeyed the order

to resume his seat without saying a word, but the vicious glance he bent upon the keeper assured me that he had to deal with a ferocious customer, who at the first opportunity would be sure to attempt an escape, by foul or by fair means.

No food had passed my lips since the morning, and nature began to demand her due in a very peremptory manner.

After preparing my simple meal and sharing it with the keeper (our prisoner refused to eat), the former proceeded to narrate the particulars of the fight in the hut.

The circumstance that only one keeper was seriously wounded in the fight was mainly due to the fact that, a few seconds before the explosion and the subsequent assault, two of the defenders of the châlet had discharged their rifles at the man who had ignited the charge.

These two shots had been fired by two of the poachers sitting on the roof, to which they had climbed by means of the smoke-hole, for the purpose of looking out and watching as much as possible the movements of the enemy. From the inside of the hut they were unable to do this, as the only window had to be barricaded for reasons of safety.

The shock of the explosion, which took place before they had time to reload their rifles, unseated and landed them on the ground outside of the hut.

This occurrence had been partly noticed by two members of the assaulting force in the blaze which followed the explosion, and these two men proceeded to seize the poachers, while the rest rushed into the hut.

After a short but sharp chase they succeeded in capturing the hindermost, who was struck down with a clubbed rifle.

The two poachers occupying the hut were standing with their cocked rifles to their cheek, when Berchtold and the rest burst into the hut.

The former, on demanding their immediate surrender, was answered by two shots ; one of them laying him low, while the second one pierced the shoulder of the keeper standing at his side.

Not content with felling two men, they clubbed their rifles, and swinging them over their heads, were about to attack the group clustering round the door, with the evident design of forcing their way out. This was, however, not to happen, for before the foremost of the two poachers had advanced a few steps he fell pierced through the lungs. His companion, who was a smaller man, had been sheltered more or less by the huge frame of his comrade ; as soon as he fell he surrendered, pitching his useless rifle into the corner.

The reader will now comprehend what a fortunate circumstance it was that the fire of two of these dare-devil fellows on the roof had been drawn, without serious results, before the moment when the assault actually took place. Had these four men retained their loaded rifles, and had they remained in the dark corner of the hut, the fight would have been of a more equal character, and the issue, if not reversed, would at least have involved a greater sacrifice of life. I passed the night, for the most part wide awake, before the fire, either watching my two dozing companions and the grotesque shadows playing about the walls, or replenishing the fire, which had to serve as our candle after the torch had burned out. Right glad I was when the gray morning light streamed in through the open doorway, and I could depart from the scene of the late fight without becoming a prey to that unpleasant feeling which undoubtedly I must have experienced had I left the previous evening, namely, that vague, uncomfortable sense of having acted inhumanly in leaving a dying man to the questionable care of his late adversary.

On reaching L—— towards noon I found that the doctor, who had been summoned from the next small town, some seven or eight miles distant, had just arrived, and held out some hope of Berchtold's ultimate recovery; though of course he would be for ever afterwards unfit for his calling as keeper. The rest were going on well. I left, L—— the next morning not a little disgusted with the heartless pleasure displayed by the villagers at the success of the keepers' raid : that a life had been victimised seemed to them as part of a just and proper punishment.

My readers may perhaps ask why the poachers did not surrender to an overwhelming force at the outset of the fight. I think I have already partially answered this question when I said that a genuine Tyrolese, reared in the secluded parts of the glorious Alps, values freedom and liberty more than life itself. This feeling, together with the fact that poachers, by their reckless daring, often succeed in vanquishing a superior number of keepers, will explain the apparent imprudence of their resistance, which, I am nearly convinced, would have brought them through, had it not been for the stratagem of the wily ' Herr Oberförster.'

The worst feature of such adventures is that scores of brave lives, gifted with powers of endurance and strength almost superhuman, are thus sacrificed ; and generally speaking it is just this vigour and force which leads its possessors astray. The poor fellow turns poacher simply for the love of that most exciting and dangerous sport, the chase of the chamois—an animal which has, indirectly, brought more lives to grief than the savage tiger of India or the royal lion of Africa.

CHAPTER VIII.

A TYROLESE SMUGGLER AND HIS LIFE.

FIVE and seventy years ago smuggling was one of the chief resources for many of the inhabitants of remote valleys and. glens in Tyrol, adjoining either Bavarian or Italian boundaries.

The Tyrolese smugglers were renowned in those days not only for the bold and cunning manner in which they carried on their dangerous trade—often on an amazingly large scale—but also for the daring courage with which they resisted the armed excisemen. Nowadays the decrease of duty on the two or three articles that were smuggled, such as tobacco and silk, into Tyrol, and gunpowder, schnapps (spirits), and salt out of it, renders it far less remunerative than formerly.

Nothing proves the decrease of smuggling more strikingly than the fact that, while formerly forty and fifty smugglers and customs officials were annually killed or severely wounded in nocturnal encounters in the by-ways of the Alps, nowadays scarcely four or five men fall victims to the rifle of the officer or of the smuggler.

Pitched battles between small bodies of the detested ' Grenzwächter,' or ' Finanzer '—customs officers—and well-armed free-traders, were of yore by no means rare occurrences ; but now, owing, as I have said, to the decrease of duty, they happen but very rarely, and no doubt the next ten years will witness the total extinction of an interesting race, that of the ' Schwärzer ' or ' free-trader.'

In speaking, therefore, of Tyrolese smugglers of the old and genuine type, hardy and daunt-less mountaineers, wily and resolute foes of the Government officers, we are speaking of beings of the past, and just on that account it may prove of some interest, perhaps, to touch upon the manifold dangers that beset the path of these daring fellows, before their existence becomes a matter of tradition, or at the best, of hearsay.

In the course of my wanderings in Tyrol, and among the queer people met in odd, out-of-the-way nooks and corners, I have come across not a few smugglers and ex-smugglers. A little practice and close watching of a man's behaviour

soon enables one to say, after a quarter or half-
an-hour's conversation, if he is or was a member
of the fraternity in question. In many instances
I have succeeded in drawing out my victim by
the dark hint that I was aware of his present
or former avocation ; and my assertion, based, I
need hardly say, upon my impression only, has
been generally rewarded by the mention of one
or two interesting adventures, told with that trusting sincerity and quaint humour, entirely free
from bravado or exaggeration, which distinguish
friendly intercourse with Tyrolese in remote
districts, when once you have known how to
gain their confidence.

The most interesting man of this stamp I
have ever met with was beyond doubt Johann
K——, whose acquaintance I happened to make
in an odd manner.

Eight or nine years ago, in fact one of the
first summers I spent in my second home,
Tyrol, I was making a pedestrian tour among the
medium-sized mountain ridges that skirt the
Achenthal, close to the Bavarian frontier. One
day, while I was yet high up on the peaks, night
overtook me, and not being acquainted with the
ground I intended to pass, and no Alp-hut being

near, I had to make the best of a small log-hut, erected by the owner of the elevated pasturage as a store-house for the winter's fodder.

On entering by the square hole, about 3 feet by 2 feet, cut in the solid timber, I found the lower partition of the hut, measuring perhaps 13 or 14 feet square, empty. A ladder leading up to a square opening in the boards that formed the ceiling invited a closer inspection of the top story, in hopes of finding a couple of armfuls of hay for a bed. The roof, shelving down on both sides, was in the centre only three feet from the floor, so that an erect position being quite out of the question, I had to crawl about in search of the hay. In one of the corners I at last came upon some spread out and flattened down by its frequently having been lain on.

Finishing the remains of a very frugal dinner, I was soon in possession of this soft corner, and shortly afterwards fell asleep with my head resting on my Rücksack.

Two or three hours might have passed, when all of a sudden I was awoke by a heavy weight bumping against my side. Lying quite still, I soon became aware that it was a man who had

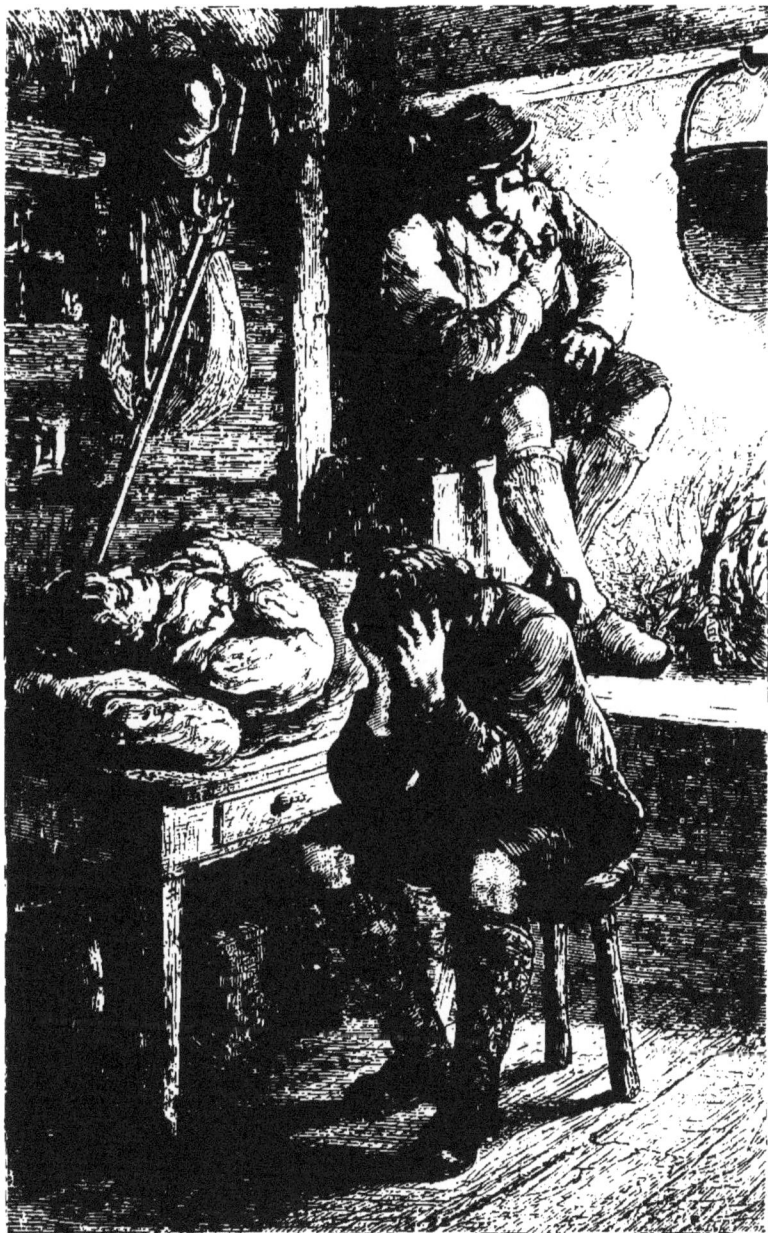

MY NIGHT-WATCH IN THE CHALET.

thus disturbed me. Five minutes later loud snoring proved that he was fast asleep.

Now only did I rise upon my knees, and creeping forward, take a peep down the hole, to which I had been attracted by the light of a fire and the loud voices of several men.

The sight that struck my eyes was odd and fantastic, forcibly reminding me of the thrilling scenes in tales of robbers and brigands, with which a boy's youthful mind is enthralled. A bright fire, burning in the centre of the hut on the bare floor, showed me five stalwart men, with soot-blackened faces, lying in various poses round the burning logs, with their rifles at their side, and six huge packages piled up against the hole, which served as doorway. No doubt was left in my mind that the occupants of the hut, whose mysterious arrival I had not heard, were smugglers, and the hut their rendezvous. The manner in which this trade was formerly carried on required that there should be a place of meeting in some remote and inaccessible part of the mountains close to the frontier. Here the smugglers would meet, the Bavarians bringing tobacco and silk stuffs, the Tyrolese schnapps, salt, or gunpowder. After settling their accounts, each man paying for what

P

he received, they again parted, the Bavarians
returning with the salt or powder, the Tyrolese
with tobacco and silk, on their backs. These
meetings occurred at certain intervals, were conducted with the greatest caution and secrecy, and
always took place at night, in order that both
parties might reach their starting-point before
daybreak.

My position, of course, was not the most
agreeable. Had I been discovered by them and
suspected of espionage, my lot might perhaps
have been a somewhat tragical finish to a pedestrian tour.

Retreating to my corner when my curiosity
was satisfied, I took up my Rücksack, and hid
it and myself in the opposite corner of the hut.

Lying down *ventre à terre*, and squeezing
myself into the angle produced by the shelving
roof and the floor, I was not only pretty safe
from discovery as long as darkness reigned around
me, but was also enabled, through a chink in the
floor, which I cautiously widened by means of my
knife, to watch the company lounging round the
fire a few feet below me. For more than two
hours did I watch the group. Merry stories,
snatches of lively songs, and tit-bits of the

last village-ball scandal, went the rounds when
once business and shop had been talked over, and
the money for the tobacco and silks brought hither
by the Bavarians paid by the Tyrolese ; the salt
and schnapps which the latter had brought being
naturally of much less value, the balance owed by
them was considerable, in one instance amounting
to more than eighty florins (8*l.*), the man in
question carrying the enormous weight of 120
German pounds, or about 150 pounds English.

It must have been some time between twelve
and one o'clock when they rose and began their
preparations for starting. One of them, running
up the ladder, poked his head through the hole
and called his sleeping companion.

A couple of grunts and an audible bump of
the head against the rafters of the roof were the
signal that my bed-fellow was leaving his somewhat confined resting-place.

On emerging from the darkness, when he reached
the bottom of the ladder, I was astonished to
perceive that he had not blackened his face, an
omission which he, however, made good by pulling
out a black mask and fastening it by strings
before his face. In the few minutes that elapsed
prior to his doing so I had ample time for a close·

scrutiny. A man of about fifty-four, of large proportions and evidently great muscular strength, he seemed to exercise a sort of command not only over his two companions, but also over the three Bavarian smugglers. Taking up his huge package on his back, and his rifle at half-cock under his arm, he made his exit through the low and narrow hole that served as a door. One of his companions had gone before him to see if the coast was clear; and on his reporting that everything was safe, the fire was raked out, the bundles taken up, and a few seconds later the hut was empty.

Just five years after this adventure I was one day sitting in the bar-room of the village of A—— drinking a glass of beer after a somewhat hot and dusty tramp of many hours on the scorched high-road leading from Tegernsee to the Achensee, when a man entered the room and sat down close to me. I knew his face ; but when and where I had seen him I could not say. I began a conversation with him, asking him point blank if he did not remember me. A sharp glance from beneath his shaggy eyebrows and a curt ' No,' was his answer. After a few more words my taciturn *vis-à-vis* rose, paid for his beer, and with a short ' B'hüt di,' for a good-bye, left the

room and the house. Asking the 'Kellnerin' if
she knew who the man was, she told me in a
mysterious sort of way that he was now a well-
to-do peasant, having once been but a poor
penniless lad ; but how he had amassed his wealth
—a man with eight or nine hundred pounds' fortune is considered rich—nobody knew ; nor could
they say why pretty Nannie, the only daughter
of a well-to-do peasant, could have married taciturn and even morose Johann, twice as old as herself. On pressing her a little further, she hinted
that people said he had been years ago a daring
smuggler, and that Nannie's father was supposed
to have been one of his comrades in this dangerous trade. She had hardly pronounced the
word 'Schwärzer'—smuggler—when the whole
scene of that night in the hovel flashed across
my mind. My curt *vis-à-vis* was none other
than my bed-fellow in the hay-loft five years
ago. A couple of months after this second
meeting I succeeded, not without some difficulty,
in making the acquaintance of Johann K——,
the rich peasant and ex-smuggler.

One evening, on returning from deer-stalking
in the forests close to Johann's house, which latter
I had made my night-quarters, on purpose to have

a quiet chat, I was sitting alone with him, in front of his house, under the broad awning of the balcony, running the whole length of the first floor, when I led the conversation to the ridge of mountains— about six hours off—the site of my first rencontre. Knowing it would be useless to endeavour to gain the confidence of my reticent host by any other means, I shortly afterwards told him that I knew what his former occupation had been, and related to him how the whole thing came to pass. Jumping up, he placed himself in front of me, and offered me his brawny palm. My bold tactics had gained the man's trust, and the reticent smuggler, evidently convinced of my sincerity by my having kept his secret, was now a grave but frank man, of that bold and firm character which, in Tyrol, is frequently hidden under a mask of suspicious moroseness, repelling the approach of strangers.

That same night, sitting in the roomy parlour, uninterrupted by wife or child, he related to me his whole life's adventures and exploits.

'My grandfather,' he began, 'and my father were both engaged in the smuggling trade between M——, my native village, in Bavaria, and Tyrol. The former, owner of an inn, chiefly confined himself to concealing the goods smuggled in by others, and selling them secretly to peasants, grocers, and innkeepers. One night a descent was made on his house by the customs-officers, and before the sacks of powder and kegs of spirits that had just been brought could be concealed in their usual hiding-place, the armed officials had effected an entrance, and my grandfather and two of his mates were taken prisoners. Condemned to a long term of imprisonment, my grandfather died before its expiration. My father, a lad of twenty at the time, leaving the management of the inn to his mother, left for Tyrol, where he found employment as cattle-driver. Detesting his country, he enlisted as a common soldier in the Tyrolese ranks on the outbreak of the French war in the last year of the last century. He fought at several battles, and in one—that of Berg Isel (1809), near Innsbrück—where less than 18,000 Tyrolese peasants routed more than 26,000 Bavarian and French troops. He distinguished himself in so marked a manner that Hofer, the Tyrolese General, made him a lieutenant on the battlefield. At one of the last engagements of that memorable war he was severely wounded, and while he lay at the point of death in a peasant's house, the news of his mother's death reached him.

' He recovered, and subsequently married the

peasant's daughter who had nursed him through his illness.

'Fearing to return to Bavaria, lest he should be prosecuted for espousing the Tyrolese cause in the late war, he sold the heavily-mortgaged inn, and dividing the proceeds with his brother, invested his share, amounting to a few hundred florins, in cattle. He made one journey to Central Russia with his breeding cows, but on his way back was robbed of every penny, and he gave up this business. As I had been born in his absence, he decided, on the earnest wish of my mother, to turn to farming. Renting a small peasant's cottage and three or four acres of land, he recommenced life. His hopes of succeeding in his farming, however, were destined to be disappointed, for hardly had he been on his farm a year when the murrain killed his two cows, and he was at starvation's door.

'In this moment of need his brother, who, it seems, had kept up a connexion with the smugglers with whom my grandfather had been associated, succeeded in enticing my father to join him and three or four other daring fellows, to establish a regular smuggling trade between Kufstein and a small townlet in Bavaria.

'The Alpine passes traversed by these intrepid

free-traders were high and steep, rendering each venture or expedition a fatiguing march of some ten or twelve hours. All went well for a year or so, till one unlucky night my father and three others were successfully waylaid by a party of six customs officials. The "Halt, or we shoot!" ringing out in the dark night at a few paces' distance, brought my father's rifle to his shoulder—he usually walked with it under his arm at half-cock—and before the aggressors had the opportunity to act upon their threat, my father had fired at the dark form of the leader, hardly five or six paces off. The path was at that point very narrow, and skirted on one side by a high wall of rock, on the other by a diminutive precipice, some twenty or five and twenty feet in depth, ending, as my father knew, in ground covered by the dense brushwood of the latschen. The moment he fired he leaped down the precipice, four or five shots passing over his head. . The weight of his load saved him, for he fell on his back, the strong wicker-work '.Kraksen,' in which he carried the gunpowder, the article of his venture on that occasion, breaking his fall.

'The man in his rear was shot, while one of the remaining two was taken prisoner, the third escaping.

'Hastily hiding his goods under some brushwood, my father took to his heels, and reached home in safety before daybreak. This unpleasant rencontre naturally cast a deep gloom over the members of the " company " [as my informer naïvely termed it]. The man who had been shot died the same night. The official whom my father had shot at was wounded in the arm, while the second member, who, as I have related, was captured, proved "game," and resolutely refused to mention the names of his comrades, though he well knew that his sentence would only be the severer by his reticence.

' Notwithstanding this, however, suspicion fell upon my father, and the house was ransacked by customs officials. Not finding anything of a suspicious nature, my father escaped with a solemn warning. For nearly two years their trade was at a stand-still, and it was only when dire want stared us in the face that my father thought of resuming his dangerous traffic.

' This time, however, he undertook it alone, and on his own account ; and by dint of great caution, and by leaving an interval of more than a week between each journey, he managed to escape detection for a considerable period. Once, indeed, he was on the point of being discovered. The man who always met him on the frontier to exchange tobacco and silks for the spirits or salt, had been prevented by some reason or other from keeping the rendez-vous.

'After waiting the whole night for him in the usual place, a cave, my father determined to pass the frontier, and repair to the man's habitation, an outlying peasant's cottage four or five hours off.

'Having washed his blackened face at a brook —as in daytime it would tend to attract attention—he secreted his rifle in the cave, and then crossed the imaginary frontier line, formed by a high ridge of mountains, and entered Bavaria, his native soil, untrodden by him for many years, though his "trade" brought him to within a few yards of its boundary forty or fifty times in the year.

'He had not proceeded far down the slopes on the Bavarian side, when he perceived, a short distance off, a Bavarian "Grenzwächter."

'Trusting he would let him pass, under the supposition that he was a peasant on a legitimate errand, and seeing that flight was impossible, he continued to walk on.

'Whether it was that some remnant of soot on

my father's face, or some other sign, roused the officer's suspicion, certain it is that on coming up to him he ordered my father to show him the contents of the " Kraksen " on his back.

' Resistance to this command, unarmed as he was, would have been madness, the official having his gun at full cock in his hands, ready to shoot at the first sign of resistance.

' My father, pulling down his Kraksen, and playing the part of a pig-headed peasant lout, replied that " he well knew that there was no law compelling a peaceful peasant, carrying his butter from his châlet to the village, to show the contents of his Kraksen to every man who might desire it. If he wanted to see what was in it he would please kindly open it himself, for he would not." The officer, though assured by my father's quiet tone that he was not a smuggler, but rather a stubborn peasant boor, thought he would punish this saucy demeanour by turning the contents of the Kraksen upside down, and laying aside his gun, bent down to unfasten the divers strings that held down the lid. This was just what my father had waited for ; and with one sledge-hammer stroke of his enormous fist he floored the unfortunate officer.

' My father of course decamped with his Kraksen,

but before doing so he broke the officer's rifle, sword, and bayonet across his knee, leaving the pieces in a pile by the side of his senseless foe. Strange to say, he never heard any more of this affair ; but he vowed that he would never again cross the Bavarian frontier, and he kept his word.

'Several years passed, and I was about fourteen, when one day my father called me aside and told me in his abrupt manner that he would take me with him on a "journey" that night. My father's manner and serious tone assured me that my accompanying him was no ordinary occurrence of life, an impression rendering superfluous the caution that I was to keep all that I might see or hear a profound secret. "If you behave well and do all that I tell you," my father continued, "you need not attend school any longer." Now this was a grand and joyous vista to a boy who detested school work as I did ; and though as five months of the year were holidays, and I was in the last year of school, my joy was perhaps foolish at my sudden promotion to manhood, yet nevertheless, that day was the happiest of my life.

'Full of impatience and curiosity, I refrained from retiring to my bed at the usual hour of eight or half-past, but waited up for the return of my father, who had gone out when he had finished his tilling for the day. I need hardly tell you that my father's occupation as smuggler had been kept a dead secret; only my mother knew of it, and when now and again I met him returning home at an early hour in the morning, I never troubled my mind about it.

'At nine o'clock my father returned, and bidding me follow him, led the way out into the dark night. For two hours he walked on with his usual quick and long step.

'We had passed up through a dense forest, and on emerging from it crossed a small plateau, on which were scattered here and there log-built huts for hay.

'The one highest up belonged to the peasant property which we rented. A low whistle of my father was answered in the same key, and we jumped through the hole giving entrance to the hut.

'By the light of a small lantern, which my father lit, I perceived three men sitting on logs. Only when two of them accosted me by my name did I recognise my uncle and one of our neighbours, their blackened faces disguising them completely. The third man was a stranger to me.

'Pulling out a box full of soot, my father proceeded to blacken his own face and mine. While we were busy, two of the men had pushed aside a heap of hay in one of the corners, and after removing a few inches of earth, they laid bare a sort of trap-door. Opening it, they both disappeared in the cavity below it, reappearing in a few seconds with two large Kraksen.

'This manœuvre they repeated twice or three times, bringing to light two more large Kraksen, a smaller one which was apparently empty, and four rifles.

'The smaller Kraksen being filled with hay, and the lid carefully bound down, my father told me to take it on my back, and proceeded to give me his instructions. According to them I was to proceed at a moderate pace up a certain path leading towards the Bavarian frontier, and passing a deserted châlet, about two or two hours and a half from our starting-point.

'On approaching this hut I was to sing a certain "Joddler." A whistle from within would be my signal to enter the hut, but before entering I was to "joddle" in a loud voice. On my way up, my father continued, I should at intervals of five minutes give the signal that all was right by singing. I may mention that I was by no means a bad singer, being not only a strong boy for my age, but possessing great taste for music, and a strong voice.

'The four men were to follow in my wake, leaving a certain distance between me and them.

'The nature of the business was now no longer a riddle to me ; and thus my father's hint, that in case I should be stopped by anybody I should desist from "joddling," and so give them a negative warning, was quite superfluous.

'A little before half-past eleven I started in my new character as scout, and right merrily did I make my " Joddlers " ring out in the dark night, the surrounding heights and precipices returning the sound two and three-fold.

'In the allotted time I reached the hut, and my merry " A braunauged Dirnd'e h'an i'im Herzen " ("A brown-eyed maid is in my heart ")—the song indicated by my father—was answered by the preconcerted low whistle. The inside of the châlet was very similar to the one I had left two or three hours ago, the only difference being that a fire was burning on the ground, round which four men were taking their ease. The single window there was boarded up so that not a ray of light would betray them, and with their rifles at their side the men were evidently prepared for danger.

' All of them being strangers to me, my position was for the first moment somewhat embarrassing.

' For the first moment, however, only; for, slapping my back, and praising my accurate observance of the instructions received from my father, they offered me a bottle of schnapps, and, after a good pull at it, the owner invited me to share his seat beside the fire. How grand it seemed to me thus to be treated as a man and fellow-smuggler! How elated I was at the few words of praise that fell from the lips of my " companions "!

' My father and his three confederates arrived shortly, and now for the first time I learnt that the venture of that night was one of especial importance, the smuggled goods being of great value. The Bavarians, for such were the first occupants of the hut, after paying for the goods and leaving their bales of tobacco, departed shortly afterwards, it being later than usual.

' Our return was performed in a similar manner to our journey thither, and having deposited our Kraksen and rifles in the usual hiding-place, we reached our respective homes shortly after break of day.

' Thus ended my momentous *début* in the character of smuggler. The sense of danger

Q

lurking at one's heels, the free life, and lastly, but
not least, the animating influence of the constant
state of alertness which must distinguish a smuggler
successful in his craft, engendered in me the resolu-
tion that henceforth free-trading should be my occupation, and success in it the goal of my ambition.

'For two years I acted as my father's scout, and
on two different occasions did my tactics save him
and his companions.   Stopped in my nocturnal
wanderings by the usual "Halt, or we shoot!" of the
" Grenzwächter," you can paint to yourself their disappointment and mortification when the supposed
smuggler turned out to be but a poor " Wurzen-
graber "—digger of roots—and the contents of my
Kraksen, the object of their researches, proved to be
roots of the Gentiana [1]— or other Alpine plants.

'My two years' apprenticeship had made me an
expert and daring smuggler, and you can conceive
my pleasure when one day my father announced to
me that henceforth I should participate in their
gains, and "carry my own goods."

'To enable me to buy the necessary stock for
my first two or three ventures, my father handed
me a comparatively ample sum of money, making

[1] These roots are used very largely for distilling purposes, a
strong and bitter spirit being manufactured from them.

me, however, promise that I would pay off my debt
by instalments.

'For two years our trade went on swimmingly
and I was laying by money for the proverbial rainy
day. Sooner than we thought did it make its appearance. One night on our return from the usual
place of meeting, as we were hurrying down the
narrow path leading to the hut where we used to
conceal our goods, the ominous challenge of the
Grenzwächter brought us to a dead halt. From
the front and from the rear we were enclosed, and
the formidable precipice at our side prevented any
escape in that direction.

'My father, who was leading, fired, I following
suit a second later. Of what happened afterwards I
can give you no clear description. A fierce struggle
with one of the Grenzwächter occupied me for the
next few minutes. My great strength enabled me
to rid myself of my foe very soon. Not so, however, of one of his mates, who, larger than I, made
a fierce rush at me the moment I had regained my
breath. I closed with him, and a terrible struggle
began. Hither and thither we swayed, both of
us trying to use our knives, but each firmly grasp-
ing the arm of the other. At last my firm grasp
with my free hand upon my foe's throat began to

tell, and a few seconds later he was lying half-dead at my feet. My father, who had shot the leader, had been himself wounded by a bullet, but not so severely as to render him *hors de combat.* One of our two confederates was disabled, the other was engaged in a fierce combat with two officials, who were endeavouring to get at him with their swords, while he kept them off with his clubbed rifle.

'Matters were terribly critical, but there was yet some chance of escape for those who were not disabled, when, to my dismay and horror, I heard shouts of approaching men, and a second or two later three shots rang out, and my father, to whose aid I was just making, fell to the ground with a groan. The feeble moonlight enabled me to perceive that a reinforcement of three men, pro-bably stationed further down the road, had arrived.

'They were standing two a-breast, the third at their rear, when, maddened by my father's fall, and knowing that this was my only chance of escape, I rushed at them, and by the mere impetus of my attack sent one sprawling to the ground, while the second gave way, and the third, at his back, was floored by a blow of my clubbed rifle. Pursuit was vain; my limbs and sinews, strung to their utmost, would have defied much fleeter men than they. I reached home covered with perspiration, and nearly out of my wits at the fate of my father. Help of any kind was out of the question, and the only thing that remained for me to do was to inform my mother of his fate, and collect such trifles as I needed, together with the money I had saved. I knew that in a few hours our house would be closely searched for me. Bidding a tearful farewell to my mother, and telling her to write to me to her brother living in South Tyrol, I was off within twenty minutes of my arrival.

'Skirting the high-roads, and keeping to forest-paths, I was fortunate enough to reach the next town, within fourteen hours of my leaving our remote homestead.

' I slept in the hay-loft of one of the houses outside of the town, and proceeded on my weary tramp the next day at sunrise.

' Eleven days of marching brought me finally to my destination, my uncle's house, where I found a letter from my mother, in which she informed me that my father had died shortly after receiving his second and fatal wound, that one of our companions was severely wounded, and the other captured.

' The Grenzwächter had two dead and three

wounded ; you see, therefore, that our resistance was a vigorous one.

'For more than five years I stopped with my uncle, aiding him in his timber trade, and extending a helping hand wherever it was needed. On my uncle's death I inherited half his modest fortune, which I embarked in cattle. In the course of the next fifteen years I made a number of journeys to Russia with varying success, so that at the end of this period, on getting tired of my wandering life, I found myself the richer by nearly 2,500 florins (less than 250*l.*). I gave up my cattle business, and being then nearly forty, I resolved to marry.

'My mother had died years ago, and the residue of my father's savings his brother had received.

'On visiting my old home I could not refrain from seeing if my smuggler comrade, who had been taken prisoner that disastrous night, was still living. On entering his house, quite close to my home, now in strange hands, I learnt that he had died ten or twelve years ago, and that his widow had married again. His daughter had accompanied her mother to her new home, some distance off, that peasant's house yonder. Having nothing better on hand, I determined to visit the widow of

the most intimate friend of my youth. On this
visit I made the acquaintance of Nanni, now my
wife. Young, very pretty, gay, and well aware
that she was the heiress to a goodly fortune for a
peasant girl, she lent anything but a willing ear to
the courting of a somewhat mysterious personage,
more than double her age (she was then seventeen),
with no home over his head, and for aught she
knew, a penniless beggar; I had refrained from
telling her or her mother of my savings. Twice I
asked her if she would have me, and twice I was
refused. Humbled in my own eyes, and mortified
at the girl's disdain, I left her dangerous neighbourhood shortly after my second repulse.

'In my frame of mind, dissatisfied as I was with
myself and with the world in general, the recollec-
tion of my youthful life as smuggler had a strange
charm; what if the mature man, long past the giddy
days of youth, should exchange a life of daily
drudgery and poor returns for the free and animating avocation to which I had served my apprenticeship twenty long years ago? More and more did
this plan attract me, and from day to day the life
of a smuggler, with its constant danger, seemed the
only way to dispel my discontent. Determined
and impulsive as I am, it did not take long to

ripen my plans. My money placed in safe hands, I at once made overtures to a set of smugglers by reputation more daring and bold than the ordinary run of men of this stamp. A week later I was a member of their "company," and had opened my campaign with an expedition of more than usual importance.

'Chopping and changing from one place to another, just where my fancy and the promise of large returns led me, I passed seven years. A lull in my trade enabled me to pay a visit to the house of Nanni's step-father. I had not seen her during the intervening years. Handsomer than she was at seventeen, sedate, and more attractive than ever, the girl enchained my heart a second time ; this time, however, my wooing was crowned with success, and a few months later I led my bride to the altar. My savings and the returns of my seven years' smuggling ventures had nearly quadrupled the original sum. I bought the house we are sitting in and the twenty-five acres surrounding it. For several years I lived the life of a steady-going peasant, happy and content. Gradually, however, my quiet, humdrum life began to pall upon me, and an irrepressible longing to return to my old life came back. Rich, with all the

comforts of life I desired, a loving and devoted
wife at my side, and two children at my knee, I
might well have been thought mad to endanger my
life by exchanging my present position for that of
a smuggler. Still do what I would, the recollec-
tions of my old life were for ever dazzling my eye.

'My former confederates, eager to win me
back to my old course, succeeded at last in their
endeavours. On and off, leaving often an interval of a month between each venture, I left
my home for the two or three days necessary to
reach and return from the scene of our smuggling
operations. Fortune seemed to favour me, for
not once were we stopped. My three companions,
who looked upon smuggling as the means of
gaining their daily bread, and not, as I did, as
a pastime, had been fortunate in their transactions,
so that one by one they dropped off, settling down
in each case as steady peasants. The time you
saw us we had lost only one member, the second
one following his example a short time afterwards.
My wife, to whom I had confided my design,
was of course greatly against it from the beginning, imploring me to desist from my ruinous
procedure. Four years ago, when my third and
last companion resolved to bid adieu to the trade,

she at last succeeded in making me promise never
again to put the mask before my face.

'Since that day I have lived a happy and con-
tented life ; the youthful fire has burnt out, and the

PORTRAIT OF 'JOHANN.'

wreck of the former smuggler is stranded high and
dry on the shore of home life.'

It was late when this simple narrative of a
life of restless adventures came to a close, and
the stalwart, broad-shouldered man of sixty, rising
from his seat, proffered me his brawny palm.
With mine resting in his strong grip, and with

glistening eyes, while pointing to the door of
the next room, where his wife lay asleep, he
remarked with deep feeling, 'My life's gratitude
cannot repay my debt to that woman ; she it was,
and she alone, that saved me, perhaps from an
ignominious death, and made me the man I am.'

CHAPTER IX.

THE BLACKCOCK.

THE Capercali, the largest of European gallinaceous birds, and the Blackcock (*Tetrao tetrix*) are the two largest game-birds of Tyrol. Both belong to the grouse species; but while the former is of gigantic size, weighing as much as from ten to fourteen pounds—in fact quite as large as a turkey—the latter is much smaller, his weight but rarely exceeding four pounds. Though the capercali is the more magnificent bird of the two, the blackcock is considered the nobler game. Far shyer and more cunning, the latter is very difficult to shoot in Tyrol, and the sport requires great hardihood, patience, and an accurate knowledge of the bird's peculiarities.

I believe these fine birds are to be found in some districts of England, specially on the estates of the Marquis of Anglesea; and from certain historical accounts it appears that both the black-

THE BLACK COCK LISTENING TO THE LOVE-SONG OF A RIVAL.

cock and the capercali were once very abundant
in the forests of Scotland, though the former had
always the privilege, and was considered 'royal
game.'

Both these species of grouse are shot· in Tyrol
on quite a different principle to that in England,
where the shooting commences on September 1.
In Tyrol, on the contrary, they are shot during the
pairing season, in April and May, the hen-birds
being carefully spared.

Strange to say, the sight and ear of the black
cock assume during the pairing period an amazing
keenness, while those of the capercali remain very
much the same throughout the year.

This of course renders blackcock shooting,
although an interesting, by no means an easy
sport. As with chamois-shooting, there are various
ways and means of making it easier, and these
are generally adopted by gentlemen who have
well-stocked preserves, and who shun the fatigues
and exposure to the cold incidental to the genuine
sport. With the increased ease much of its charm
vanishes, and to speak candidly, I would rather shoot
one cock according to the regular Tyrolese fashion,
alone and unaided by any artificial contrivance,
than half-a-dozen from the hut erected near the

tree where, for days previously, a cock has been spotted by a keeper. I must add that the black-cock, if he remains undisturbed, invariably returns every morning from his haunts lower down in the woods, during the whole of the pairing season, to one and the same tree, perched upon one of the branches of which he sings his lovesong. It is therefore not difficult, when once a cock has been spotted by a keeper, and a miniature hut has been run up in the course of the day close to the tree in question, for the noble master to slay his royal game. It is simply a question of sitting a few hours, well wrapped up in coats or furs, patiently awaiting the advent of the game. Far different from this is the genuine sport. An account of an expedition of this kind may give some idea of its attractiveness, though perhaps but few would be willing to share the fatigues and exposure to cold incidental to it.

The difficulties of its pursuit in the pairing season are much enhanced by the great elevation of the spot selected by the cock for the scene of his amorous adventures, and of the fierce combats which generally precede them. I have known as many as three or four fights take place before the cock, who proves himself victor over his two

or three rivals, can commence his strange antics
and odd-sounding lovesong, for the edification of
the hens who crowd round their polygamous lord
and master. Nothing is more ludicrous than to
see the love-sick cock, full dressed in the glory of
his glossy steel-blue plumage, strut round the
base of the tree selected for the scene of action.
Now trailing his wings, turkey fashion, and inflating
his glistening throat ; now throwing back his head,
his neck waving to and fro, while the tail is ex-
panded to its full, standing at right angles to his
body ; then again, in the ecstasy of passion,
trembling all over his body, while froth issues from
his beak, and the eyes are covered with the nicti-
tating and glittering membrane, he will gambol
and throw some somersaults with amazing rapidity.

The lovesong of the cock is, strange as it may
seem, a matter of great importance to the sportsman. It consists of three distinct notes, or 'Gsatzln,'
which are repeated constantly, and at more or less
regular intervals. Resembling the lovesong of
the capercali, though much louder, the first and
second notes could be compared to gurgling chuckles,
while the third, 'das Schleifen,' might be compared to the sound caused by sharpening an edged
tool on a whetstone. The third note is the one for

which the sportsman must wait. During its utterance the cock is entirely insensible to danger ; his passion in this second or two is so excessive that sight as well as hearing are dead to all other influences. While it is being repeated the hunter may advance, and can even fire off his gun without disturbing the bird ; while during the two first notes, and during the intervals, the most perfect silence must be observed by the hunter, hidden by rock or brushwood from the amazingly keen sight of his game. A suppressed sigh at a distance of many yards is sufficient to send off the alarmed cock.

But now to my own account of a blackcock-shooting expedition. With a pair of snow-hoops, my trusty crampons, and a single-barrelled large-bore fowling-piece, and with my usual bag, filled with provisions for three or four days, on my back, I started on a fine April morning for the scene of action, a remote valley some eight hours off. A week's bright sunshine had melted the snow on my path, and even for several hundred feet above me the Alpine pasturages and sombre, dark-green pine-forests clothing the adjacent slopes were free of their white pall. Arriving in due time at a small peasant's cottage—the last house on my way—I

determined to remain there till fall of night.
Entering the general room of the house, I received
a warm welcome by its owner, his family, and
Lois, a daring young native sportsman, who had
often been my companion on shooting expeditions.
The rest of the afternoon and the evening—I had
decided to put off my departure till nine o'clock
at night—were passed in agreeable company,
chatting and laughing over our glasses of schnapps,
that being the only liquor the man had in his
house. A number of forgotten adventures and
odd shooting anecdotes, in which either or both
of us had played a part, came upon the *tapis*,
to the great mirth of the whole party, so that
when the crazy old clock in the corner of the
wainscoted room began to 'hum and haw' preceding the final effort of striking the necessary
nine strokes, I was sorry to be obliged to leave
the merry company, and exchange the cosy warm
room for the bitterly cold air outside. On issuing
forth, we saw the full disc of the moon just crest-
ing the high ridge of snowy mountains, at the very
base of which lay the narrow glen in which the
cottage was situated. The cold, although it was
the latter half of April, was intense; but I was
very soon, by dint of fast walking, in that pleasant

R

state of warmth peculiar to violent exertion in cold weather. Putting my best foot forward, I had within five or ten minutes reached the snow-line again. Fastening the snow-hoops to my feet I began work in earnest. As I sank nearly up to my thighs at every step, it took me more than three tedious hours to gain the first eminence, some two or three thousand feet over the hut. The dry, powdery state of the snow had gradually given way to a firmer substance, and at last, on reaching the top of the ridge, I found the snow 'harscht,' or frozen. Owing to the depth of the ravine up which I had traced my steps, the rays of the sun never touched its sides, and the snow was therefore powdery and unresisting ; higher up, on the contrary, the sun had melted the top layer of snow, which, in the long hours of the night, froze, and resembled as much as possible the smooth surface of a glacier after a hot August sun has polished it. My snow-hoops now of course became not only useless, but actually dangerous. Unfastening them, I strapped my crampons on and got my small ice-axe ready.

The moon shining brightly, night was changed into day ; it was therefore easy to continue my way up the next ridge, from the base of which

I was, however, yet some little distance off, a sort of
miniature valley lying between me and the point
where an ascent up the very precipitous slopes
was practicable. Well acquainted with the terrain,
I knew there was no chasm or rocks at the bottom
of the gully, and imagined there was no danger
attendant on sliding *à la Tyrolese* down the icy
slope which, as I have said, I had to cross.
Cutting two or three pine-branches off the next
tree, I entwined them so that they should furnish
a sort of seat. On this I sat down, and digging
my ice-axe, as a sort of drag, into the glistening
surface, I began my descent. As the slope was not
very steep at first, my drag was of sufficient re-
sisting power to check the pace; but soon, to my
dismay, the gradient grew steeper and steeper,
increasing in a proportionate degree the speed at
which I was travelling. My axe was wrenched out
of my hand, and I was left to the mercy of the
hindermost spokes in my crampons; but these,
owing to the position of my body and my feet, only
scratched the ice, checking the speed but little. The
slope was some 900 or 1,000 yards in length, and
before I had reached the middle even this mode of
checking my downward course became too dangerous to continue; for had my crampons come in

contact with the slightest unevenness, or with the smallest stone imbedded in the ice, I should have been jerked head foremost off my seat, and left to continue my course at lightning speed in any but a comfortable position. Fortunately this did not occur, and I reached the bottom of the gully seated on my primitive sledge. Though my whole downward slide could not have taken more than four or five seconds, the terrific speed had taken away my breath, and, what was worse, the impetus had driven me far into a snow-drift of large dimensions, which had accumulated at the foot of the slope, and which, as it was under the lee of a high wall of rock, was protected from the sun, and consisted, therefore, of powdery loose snow, offering hardly any resistance to my mad onslaught, which carried me right to the centre of the huge hill. After working myself out and dusting my coat and trousers (my gun-lock was protected by a mackintosh wrapper), I started once more up a steep incline covered with a coat of ice, or rather frozen snow, polished and smoothened by the action of a warm April sun and intense cold at night. By two o'clock in the morning I reached the top of the mountain, or what might pass for it, the scene of action. I have said that the fact of

knowing the precise spot where a blackcock holds
his love court facilitates to a great extent the
final result. Now the ridge of mountains upon
which I was standing was some three or four hours
in length, and probably along the whole of it not
more than one, or at the utmost two, blackcocks
could be found. The choice of the right spot thus
became a matter of luck. To some extent, of course,
one can be guided in one's selection of the spot one
intends to watch by the fact that they generally
choose the very highest points of the mountains,
selecting, if possible, for their head-quarters an old,
gnarled, weather-beaten pine, or 'Zirbe,' a species
of pine growing only in the highest regions of
vegetation.

By the time I had eaten a piece of bread and a
small bit of bacon, swallowed a gulp of the 'Enzian
schnapps,' and turned over in my mind the various
'Stände' on that ridge where a cock could possibly
be, it was close upon three o'clock, and therefore
the very best time to proceed to the spot selected.
The moon had disappeared, and I was glad I had
no very bad places to cross on my way to the spot
chosen by me as the most likely, if not for seeing
a cock, yet at least for hearing him and so spotting
him for the next morning.

A quarter of an hour's cautious climbing brought me to the northern extremity of the ridge, where in gigantic steps of a couple of thousand feet each, the mountain abruptly fell off down to the valley, some four or five thousand feet below me.

Quite close to the spot where I had killed a fine cock the year before, I hid myself as much as possible behind the tough branches of a Latschen bush, about ten paces from a huge patriarchal 'Zirbe,' stripped of nearly all its branches, by repeated strokes of lightning, and rearing its gaunt, gnarled trunk into the starlit sky. For the next hour all was silent round me ; the intense cold, abetted by a piercing wind, succeeded in making my place of ambush as uncomfortable as possible. Shortly after four o'clock the heaven began to show signs of approaching day. The snowy peaks which reared their noble forms all round me were one by one lit up with the exquisitely rosy tint peculiar to the reflexion of the earliest rays of the sun on unbroken surfaces of snow. As yet the sun was not up, and would not be up for at least a quarter of an hour ; in fact it was just that moment when the blackcock, whose maxim is 'early to bed and early to rise,' shows the first signs of life.

A distinct 'whirr' close over my head told me

that my selection had been a good one. Hardly daring to look at the tree, for fear of betraying myself to the cock, I perceived, relieved against the light sky, the noble bird seated on one of the remaining branches of the Zirbe-tree.

I could do nothing, not even raise my gun, till the third note of the song assured me that the cock was at the height of his passion. A flap of his powerful wings, and he had changed his perch to another branch higher up, but hidden from my view by the trunk of the tree. The next minute the love-sick cock was singing. Was I to wait till he flew to the ground and began his amusing antics, running the chance of losing him out of sight ; or was I to endeavour to 'anspringen,' the process of gradually approaching him by a series of jumps or strides, performed while the cock is singing the third notes? On the other hand, delay seemed imprudent, as by his song I knew the cock to be an 'old' one—that is, three years of age—and therefore of a particularly jealous disposition, eager to fight any young interloper who might betray his presence in the old cock's preserves by singing. As, further, it was very early in the season, and thus it was likely that the cock had not yet settled down to any one definite spot for his morning song, but was shifting

about from place to place, singing a few stanzas at
each, I presumed it was the safest course to try
' anspringen,' consisting in this instance of shifting
my position a little to one side, in order to get a view
of the bird. On my right, not more than a foot, an
immense precipice fell off, so in order to hide myself
I had to move to the left, over some rocks bare of any
vegetation. *Ventre à terre*, I awaited the signal to
move, namely the third note ; then jumping up and
running forward two or three steps, I had at the
conclusion of the third note, which lasts but a few
seconds, to throw myself down again, remaining
quite motionless till the next ' Gsatzl.'

Three of these momentary but frantic leaps
brought me to the desired spot, from whence I
had a full view of the cock, and the very next
' Gsatzl ' of the bird was intended by me to be its
last.

Luck, however, forsook me at that moment.
Inflating his throat and expanding his magnificent
tail to its full, he was just about to commence the
second note of his dirge, in my full view, hardly
thirty yards off, when with a slight crack a small
twig snapped asunder under my weight. The next
second, before I had time to raise my gun to venture a flying shot, the cock was off, passing in his

short but 'dipping' flight the very bush behind which I was hidden.

Cramped with the cold, wet through from lying on the snow, and out of humour, I was just considering what to do next, when from afar, but still on the same ridge of mountains I heard the song of a second cock. The distance was too great to hold out any hopes of reaching the cock before he was off from his rendezvous. I therefore determined to 'spot' him if possible, in order that I might be sure of him the next morning.

I proceeded therefore with all despatch in the direction of the sound, and within three-quarters of an hour had reached a prominent crag, from the top of which I had a full view of the place where I supposed the game to be. Lying at full length on the eminence, telescope in hand, I scanned the isolated gnarled old pines and 'Zirben' which dotted a large expanse of barren ground, upon which, scattered about in every direction, lay huge boulders of rock. All was silent, but shortly I saw two hens take wing from beneath one of the trees some eight or nine hundred yards off. Presently the cock followed suit ; but as it was early in the season, he took a different direction, and finally, after alighting for a moment on a tree, crossed the valley at my feet, and disappeared in the morning mist that filled it.

After remaining upwards of an hour seated on my Rücksack, enjoying the splendid view rolled out at my feet, I descended to an Alp-hut half-an-hour's walk from the point I was occupying. In this hut I intended to stop during the day and the better part of the next night, leaving it an hour or two before sunrise next morning for the tree upon which I had spotted the last cock. On reaching the hut, occupying a sort of sink in the ground, I found only the roof projecting from the snow. As ingress by the door was well-nigh impossible, save by digging a cutting down to it, I preferred the other way of effecting an entrance, viz., by removing two or three of the 'Schindeln,' small boards of larch-wood, with which these huts are roofed, each board being nailed down and, further, to prevent the whole roof being carried off by the high winds, weighted by heavy stones.

Five minutes' work and a jump down the dark space landed me safely in the front part of the hut, containing a fireplace, an iron pan, a brass spoon, and a cot filled with hay. Well provided with provisions, and even the luxury of some newspapers

to pass the time, and a candle whereby to read
them, I expected—to use an American phrase—
to have a good time in my solitary habitation.
The first quarter of an hour saw a bright fire on
the open hearth, a pan full of 'Schmarn,' my coat
and boots hung up to dry, and an invigorating gulp
of schnapps going down my throat. Having
despatched a hearty breakfast, and piled several
logs on the fire, I turned in to have five or six
hours of sleep. Buried in a pile of fragrant hay, I
was as comfortably bedded as a tired man need
wish to be.

Awaking refreshed after nearly eight hours of
rest, I passed the remainder of the day and the
evening in cooking a repetition of my breakfast
for my dinner, and with reading comfortably,
stretched out on the seat running round the fire,
two or three numbers of the 'Saturday Review.'
The intellectual as well as the bodily man being in
a state of repletion, I turned over on the bench,
and the next minute I was sleeping. Long before
it was time to depart I started up with an uneasy
feeling of having overslept the right hour. Consulting my watch, I found it had stopped, so naught
remained but to climb up to my air-hole and have a
look at the moon, by the position of which in the

heavens I knew I could tell the time to within half
an hour.

Reassured, I returned to the fireplace, relit the
fire, and proceeded to brew myself a strong panful
of tea, which was followed by a 'Schmarn' and a
slice of bacon.

At about half-past two I collected my traps,
stowed them, 'Saturday Review,' candle, tea, and
bacon, away in my Rücksack, put a fresh cap on
my gun, and was just creeping out of the hole in
the roof, when my attention was attracted to a
small animal scampering away from the hut over
the moonlit, glittering snow. Guessing it to be a
pine-marten, I fired at it. My position at the
moment of firing was a somewhat critical one.
As I was balancing myself with one foot on a thin
spar inside the roof the least shock was sufficient
to knock me down from my nicely-poised post.
A heavy charge in the gun, and a proportionately
strong recoil, sent me head over heels down into
the hay, some five or six feet below me.

Reascending, I saw that the marten had also
fallen, though, as its motionless position indicated,
its fall was attended by more fatal results than my
own tumble. Creeping out, I closed the hole, and
going over to my prey, I found it to be a fine male

pine-marten, a species prized for its fur. If it be shot in winter, the fur generally fetches some ten or twelve florins (1*l.*, or 1*l.* 4*s.*). My sportsman reader will perhaps learn with surprise that I ventured to fire so near the spot where I intended to watch for the blackcock. Considering, however, that it lay on the other side of the ridge, and that the birds always roost in woods or brushwood considerably lower down, I was not afraid of any bad results. I was soon at the place of ambush selected by me the previous morning. A cold hour followed, and then the ' whirr' of the approaching cock. It was as yet too dark to shoot, for the moon had gone down some time before, so I waited patiently till break of day. Meanwhile the bird had begun to sing, flying to the ground now and again, and performing his amusing antics, of which, however, I saw but little. Again he was up on the branch, giving me a full view of his noble shape, drawn in sharp outlines on the cloudless sky. The next 'Gsatzl' saw me raise my gun, and the next second the noble bird was lying on the snow.

A far-echoing 'Juchezer,' blended with the rolling echoes of my shot, rent the air, while with a few strides I was at the side of my game.

Pleasant it is to look back to such moments as

these. The fatigues and privations which one undergoes—though in this instance the latter were not worth speaking of—only increase the exhilaration at having succeeded in spite of cold, snow, the difficulties of ascent, and all the other hindrances which obstruct the sportsman's path in Tyrol.

Far different, indeed, are the feelings of the unsuccessful hunter, returning home, perhaps after two or three days of fatigue and exposure, in the character of a 'Schneider' (tailor), the nickname given to sportsmen returning with empty Rücksack. Dejected, sullen, and disgusted, he returns crestfallen homewards. Doubly long, fearfully steep, and strangely unpicturesque and tame do the path and the surrounding scenery appear to him, while the cold or the heat, as the case may be, seems unbearable.

CHAPTER X.

A WINTER ASCENT OF THE GROSS GLOCKNER.

AMONG the manifold descriptions and recitals of travels and tours in Tyrol there are none that deal with the country and its features during winter time.

Travellers visiting the country in the full tide of sunshine and warmth have, I am afraid, very little conception of what it is like in the rough season of the year, and still less idea of the terrible straits in which the frugal inhabitants are involved by a three to five feet-high fall of snow for four and five months of the year.

I have frequently been amused to observe the curling lip and half scornful smile of some native as he watched the abortive attempt of a shivering tourist on a wet day in July or August to seek shelter and warmth in the ample folds of a shawl or great coat ; and considering that this very same man, attired in the very same garb, short leathers and frieze coat, will brave a cold of the intensity of which we in England can form no conception, the scornful derision at the effeminate stranger may well be understood. In those parts of Tyrol north of the vast mountain chain which divides the country into halves, winter lasts for many months; indeed, to speak more definitely, the fact may be mentioned that in the courtyard of Castle Matzen snow lay from November 13, 1874, till the first week of the following May. Many valleys are entirely cut off from the world, every communication being stopped by the depth of snow on the paths and roads that connect them with the next large village or town.

On the mountains the snow accumulates to an astonishing depth, masses twelve and fifteen feet being by no means unusual; and Alp-huts situated a few thousand feet above the base of the adjacent valley disappear entirely.

A short time ago I was one of a party of about twenty men that were called together to aid an old couple whose hut had been entirely buried by snow. After a terribly fatiguing march up slopes, which, owing to their steepness, were covered by three or four feet of snow only, we reached the site of the hut ; nothing but a gable of the roof showed that we were standing right over it.

A trench dug down to the door enabled us at last to deliver the old people, who had been thus imprisoned for nine days. Fortunately they had a goat in their hut and a few loaves of bread in their store-room ; without these they would have perished by starvation long before our arms and shovels could have liberated them from their living grave.

Two incidents of my own experience will illustrate the difficulties attendant upon winter-sport in a severe winter in the Tyrol :—the first a shooting adventure in a remote Tyrolese valley well stocked with game ; the second an ascent of one of the highest mountain peaks in mid-winter.

The autumn of 1874 was, as those of my readers who happened to be on the Continent at that period will undoubtedly recollect, a remarkably fine one.

On November the 8th, with ten companions, natives of the B —— valley, in North Tyrol, I started on a sporting expedition, intending to be away five or six days.

Our goal was a remote little Alpine ravine sur-

rounded by high peaks, affording the very best
sport possible. As our quarters we chose one of
those odd 'Wurzenhütten'—a small châlet where in
summer time spirits are distilled from the fragrant
herbs (especially the gentiana) that grow on the
slopes and rocks. This hut, about 6,000 feet over
the level of the sea, is one of the highest-situated
of the kind I know, and for its remote position, the
fact that we had a ten hours' march to it from the
last human habitation will speak for itself. We
of course expected to find the hut untenanted, the
season being so very far advanced ; what was
therefore our surprise on reaching the châlet to
find it inhabited by the young daughter of the old
rascal who was owner of this illicit distillery !

I must mention that the reason of its inacces-
sibility is to be found in the excise laws of Austria.
All spirits are subject to a heavy duty, and the
purpose of the owners of these secret distilleries
is, of course, simply to defraud Government.[1] Lena
(the daughter) had been obliged to remain 'on
high,' in order to finish a certain quantity of spirits
ordered by the innkeeper of her native village.

[1] The quantities produced in these distilleries are very small,
some distilleries averaging not more than ten or twelve gallons per
annum.

The first four days were warm and balmy, and our sport capital ; five chamois, four roedeer, and three splendid harts rewarded our pains. The fifth day, November 13th, the weather changed, and snow began to fall in such masses that on the eve of the third day we found, on our return to the hut, just the roof-beams sticking out of the snow. Lena, our cook, was glad to see daylight again, when, after some considerable trouble, we managed to dig a sort of cutting down to the door. Our bag had now increased to twenty-five head in all, nine chamois, six roedeer, and ten harts.

The snow still continued to fall, and owing to the difficulties of the previous day we decided to remain within our hut, and not to venture out into the wilderness of snow. Every three or four hours two or three of us took turns with the spade, which we had fortunately discovered in the hut, to keep open our passage in front of the door. A pack of terribly greasy cards and an ample store of tobacco and spirits helped to while away that long day ; the next was no better, the third just the same, and at last, on the morning of the fourth, the sky cleared, and it ceased snowing

To return to the village was, until the snow should be settled down, an impossibility.

Shooting was likewise impracticable, and so we had simply to wait till the cold rendered the snow more capable of sustaining the weight of a man with snowhoops. With the latter we were unprovided, never imagining that such a terrific fall of snow would imprison us. With a little patience, a sharp knife, a bit of string and cord, and the tough branches of the fir-tree, we managed to manufacture service-able substitutes, so that at the end of six more days we started, and after a most fatiguing march of nearly twenty hours we reached the snowed-up village.

Lena, with admirable fortitude and a remark-able degree of endurance, kept up with us in good style, though of course she had the benefit of our steps, or rather knee-deep holes in the snow, she bringing up the rear of our long file.

The lighter head of game, such as roe and chamois, we carried along with us ; the rest we buried in the snow.

On arriving at the village late at night, we found everybody in commotion, and full of anxiety on our account. On the morrow they had intended to send a large body of men to our aid. Our absence of more than seventeen days, coupled with the amazingly heavy fall of snow, had made the villagers fear some accident might have befallen us.

Lena, in her short leather breeches—she had donned a pair of her father's, which had been left in the hut, so as to be able to walk unhampered by the skirts of her dress—created quite a stir; and indeed the poor girl, dead with fatigue, well deserved the warm praise and the hearty shake of many a brawny palm extended to her in recognition of her brave spirit.

A week afterwards twenty-one young fellows, armed with shovels and snow-hoops, returned to the hut to fetch the ten stags still buried in the snow.

I was unfortunately unable to accompany them, but saw some of them a few days after their return.

Sleighs being impracticable, the men had to carry the stags on their shoulders, and, amazing as it may seem, there were three or four among the lot who each carried a stag for nearly an hour at a a time. As the weight of a hart showing eight or ten points is considerably more than three hundredweight, this may serve to show the powerful build and great strength of some of the inhabitants of remote valleys.

The 'Ortler Spitze' and the 'Gross Glockner' are the two highest mountains in Tyrol. Both close upon 13,000 feet, the latter was formerly supposed to be the loftier of the two; but lately, owing to more accurate measurements, the Ortler has been found to be a hundred feet higher. Though of a greater height, the latter is not nearly so noble a peak. Not unlike the Matterhorn, the Glockner is from several points of view even of a sharper and more needle-like formation.

Several ascents of this peak in the summer months—the Glockner is by no means a difficult mountain, and even ladies have ascended it—developed in me the wish to try once an ascent in the depth of winter; and though I frequently thought of this plan for several consecutive years, I never had the opportunity or time to carry it into execution.

At last, in, December 1874, I resolved to take advantage of a fortnight's spare time and try the ascent [1] I had determined upon years ago.

From what I knew of the peak I came to the conclusion that any attempt must be made from Kals; there being two points from whence this peak can be ascended, Kals and Heiligen Blut.

---

[1] See the 'Alpine Journal,' May 1875.

To Thomas Groder, the head of the guides at Kals, a man of great experience in all matters connected with mountaineering, I expressed my desire to receive accurate information respecting the depth of snow, state of the latter—if yet soft, or already coated with a crust of ice.

The answer I received was certainly not encouraging : snow nearly five feet deep in the valley, very soft, and the probability that no guide would venture to undertake so perilous an attempt.

Not easily daunted, I determined to convince myself by eyesight of the real state of things. A railway journey of ten hours—we were snowed up twice—brought me to Lienz, in the Pusterthal. Engaging a sleigh, I proceeded to the 'Huben,' a comfortable inn on the road from Lienz to Windish Matrei, at the point where the valley in which Kals is situated branches off. My coachman laughed right in my face when I answered his question, what brought me in the depth of winter, and of so severe a winter too, into the valley of Matrei, by telling him that I intended to ascend the Gross Glockner. 'Why that is beyond what a mad Englishman would do,' exclaimed the astonished native, little imagining he was in reality addressing a member of the mad 'Engländer Nation.' 'Why

look only at the eight-feet-high wall of snow'—lining
the road, cleared by means of a huge snow plough
drawn by twelve horses—'and imagine what must be
the depth of the snow high up yonder mountains ;
and they are about a third of the Gross Glockner's
height.'

Indeed the aspect of things was anything but
promising, and my driver's gloomy prophecy did
not tend to brighten my hopes.

At the inn I discharged the sleigh, intending to
stop the night there, and proceed next morning
on foot to Kals. I ordered my supper to be
brought into the bar-room in order to indulge in a
chat with mine host, whom I knew from former
times. Even he, who, I felt sure, had a high
opinion of my mountaineering experience, thought
me demented to venture on such a trip. ' In other
winters there might be a chance of succeeding, but
this year will be an unprecedentedly severe one ;
you have not a shadow of a chance to reach even
a height of 8,000 feet.'

Resolved upon trying what perseverance in a
good cause could accomplish, I started next
morning at an early hour for Kals.

A four hours' tough struggle with snow which
had fallen to a depth of nearly a foot on the path

made in the deep snow the day before by the villagers passing to and from the larger Matrei valley, brought me to my destination.

The greater part of the afternoon of that day, December 29, was spent in serious consultation with several guides, chiefly with Groder, their head. The verdict was unanimous: 'Impossible; but if you will pay us well we will try how far we can get up on the slopes of the Gross Glockner.'

Now to try and not succeed did not suit my plans at all. I told them, however, that I was willing to enter upon 'their' proposition, and would engage all such men as would volunteer, and who had had some practice in battling with snow, as chamois-stalkers. I left them twenty-four hours to consider 'my' proposition, and at their termination four men offered themselves for the dangerous work.

It continued snowing on the 30th and on the forenoon of the 31st December.

On New Year's Eve, towards dusk, the wind changed and the weather cleared, so that when I went out in the open air in front of the house, a few minutes before midnight, in order to hear them ring in the New Year, the stars were shining brightly, and the thermometer, my constant companion in

those anxious days, was marking 11° R. (or 5° F). I
returned to bed full of hope that the next day
would witness our departure, but sorry that my
favourite project of reaching the top of the Giant's
Peak on New Year's Day had become impossible,
not only on account of the unpropitious state of the
weather on the morning of the 31st, but also owing
to the religious scruples of my four guides, who
refused to be absent from the morning service on
New Year's Day.

The tolling bells and the bright sun shining
into my comfortable wainscotted chamber woke
me at eight o'clock. Looking out of the window,
which I had to open to be able to see anything, my
joy can be imagined at seeing a bright sky and a
further retreat of the quicksilver (hung up in a
shady corner) ; it now marked 12° R., thus rendering it very probable that the snow would be
in that state termed by the natives ' harscht,' able to
bear a man's weight, spread, as it would be, over
the broad surface covered by the snow-hoop.

After their dinner or, in other words, at half-past
eleven in the forenoon, we met for a final consultation in the crowded bar-room of the Wirthshaus.
We five were determined to start, however strong
and vociferous might be the party opposed to the

whole undertaking. With the words 'Hinein kön-
nen wir nur oin mal,' or 'Die we can but once,'
the leader of my little intrepid party, Peter Groder,
closed the consultation, and they all left for their
several homes, to change their dress and bid good-
bye to their families. The provisions, four bottles
of wine, two bottles of schnapps, three of cold tea,
some lard, flour, sugar, salt, six loaves of bread, tea
and coffee, were all collected on the centre table of
the room.

At one o'clock the men returned, and we set
about dividing the stores into five equal parts. I was
determined to carry my own share, and in fact, by
taking upon myself an accurate fifth part of all
danger, work, and fatigue, not to give the men a
chance of turning upon me with the excuse that
they carried more than I did, or that I took the
lazy man's post at the rear of the party.

Punctually at two in the afternoon we started.
Our aspect, wending our steps in single file through
the narrow cutting in the deep mass of snow that lay
between the houses of the village, must have been
extremely comical.

A fool's errand it seemed from the beginning to
the greater part of the villagers, but never more so
than now. Each man bore on his back an ample

Rücksack, from which dangled on one side the large snow-hoops, from the other a pair of crampons, while a short axe, or large bundles of dry wood, or the handle of a gigantic iron pan, or coil of rope, were the visible contents of the several bags as we passed the criticising review of numerous groups of natives and guides, who had turned out to witness our departure.

For nearly an hour and a half we found a comfortable path connecting the outlying peasant houses with the village.

At the last house we halted for a moment, strapped the snowhoops to our feet, and began work in earnest. Contrary to our expectations, we found the snow in the very worst state. Fine-grained and dust-like, it did not resist our weight in the very least, and when, at the outset, I saw my front-man sink in up to his thighs, my hopes grew faint, and I heard several very distinct grumbling sounds from the three men walking in my rear.

We ploughed on, however, doing our duty in a manful and spirited way. Every quarter of an hour we changed leaders, the latter of course having comparatively the most fatiguing work, making the steps for his companions.

At five or half-past darkness set in, and lighting

our two large lanterns, we continued our march by their light.

At nine o'clock or thereabouts, we reached the 'Jörgenhut,' a châlet tenanted in summer by a herd and his cattle, and of late years but rarely used by mountaineers as their night-quarters, the comfortable 'Stüdlhütte,' two hours further up, being a far more preferable abode for a night.

We halted, and digging a sort of passage to the doorway—the snow reached up to the rafters of the hut—we entered the desolate habitation. Here we intended to leave the bulk of our various utensils not actually required in the ascent.

After some trouble we lit a fire with the wood we had brought with us, and half an hour later we were sitting round a gigantic pan, filled to the brim with 'Schmarn,' and a large iron pot full of strong tea.

We had determined to try the ascent by a route entirely impracticable in summer; and as the Jörgenhut was the last Alp-hut on our way, it would be our last meal till we returned. No wonder, we sat nearly two hours over our supper, making it necessary, in fact, to cook a second edition of the 'Schmarn' and to make a third and fourth jorum of tea.

At midnight we started, leaving everything

behind save some bread, meat, a bottle of schnapps, one of tea and one of wine, and the implements, such as ropes, crampons, etc., necessary for the ascent itself. The night was one of intense cold, the thermometer on leaving the hut marked 17° R., or 6° degrees below 0° F.

For two hours our road lay along a small valley; at the end very steep slopes ensued, terminating in the large Kodnitz glacier, forming a sort of slightly inclined plateau. At the extreme end of it, in one bold sweep of more than 4,000 feet, rises the noble Gross Glockner itself.

On reaching the slopes leading to the glacier we changed our respective positions, leaving a space of some thirty yards between each of us. The first man, the centre man, and the rear man were supplied each with a lantern. The great danger of avalanches, set into motion frequently by the mere vibration of the air resulting from a shot or loud shout, made great precaution necessary.

Peter Groder, to whom I had given the command of the party, and who was by far the best man of the guides, had had the misfortune to get into avalanches twice in his life, but was saved on both occasions by a miracle.

We had been ascending the slope for about an

hour or so, when suddenly the solemn stillness reigning around us was broken by a rumbling sound, increasing in intensity from second to second, and making the very earth shake and tremble. A huge avalanche, measuring some hundreds of yards in breadth and thirty or forty feet in depth, thundered down the adjacent slopes, in unpleasant proximity to the place on which we were standing. I was just then the leading man, and on looking back towards Peter, who was walking at my rear, I perceived him and his three companions engaged in a whispered consultation. Turning, I learnt on my approach that Peter, unhinged and frightened, was endeavouring to prevail upon the others to turn back. It cost me ten minutes' talk to persuade him to continue the ascent. Silently, not daring to speak a loud word, we climbed on, now sinking up to our chests in heaps of drifted snow, now traversing the firm pathway of an avalanche, only to sink in far over our knees on leaving the track of our dangerous foe.

Two more avalanches passed us that night, and each time Groder, daring and bold as he was on all other occasions of danger, evinced signs of fear, and but for my arguments he would have turned back each time.

At half-past three we reached the glacier, and traversing its breadth, we came to another bit of stiffish climbing. At half-past six or seven we were standing on the top of a narrow ridge, the 'Adlersruhe,' that connects the Gross Glockner with some minor peaks on its right.

Here we saw the sun rise—a spectacle of unique grandeur. The cold had abated, but the wind, terribly keen, was sufficient to freeze the marrow in our bones.

On looking towards the mountain which rose in a fearfully steep incline from the point we were occupying, we perceived by the rays of the sun that the whole grand peak was one mass of pure ice.

Unfortunately we had never thought of this possibility, and had therefore failed to provide ourselves with ice-axes. The men, amazed to find ice, were for the first moment quite thunderstruck—indeed my own feelings were very much of the same tenor as those of my four guides. Fastening ourselves together with the rope, and leaving the lanterns and snow-hoops behind us, we determined to try at least what could be done with the aid of the iron shovel and the sharp and long-pronged 'Alpenstöcke' and crampons on our feet.

Hard and dangerous work it proved to be, and had we only had an axe we should have reached our goal (not more than 2,000 feet over our heads), at least an hour and a half or two hours sooner.

Cutting steps with an iron shovel into hard ice on a very steep incline, while the wind, cold and piercing, was blowing big guns, was no very inviting occupation.

The top of the peak is divided by a sort of incision—the Saddle—into two distinct horns, one, the Gross Glockner, about a hundred feet higher than the other, the Klein Glockner, which latter we had to pass on our way to the former. At half-past nine we were standing on the top of the lower horn, and there came across a phenomenon which had never been witnessed by any of us five.

The top of the Klein Glockner is ordinarily a mere sharp, knife-like edge running towards the more elevated peak, and divided from it, as I have said, by the Saddle. Instead of this we found on reaching the top that we were standing on a broad platform some sixty feet long, and from twelve to sixteen feet wide.

I was at that moment the second in the file, and sticking my Bergstock—a stout ash pole seven feet long—into the half-frozen snow, which

T

formed the platform, I found that it penetrated, and would have slipped through had I not held it firmly. On looking down through the hole which I had made with the Alpenstock I perceived, perpendicularly, some 4,000 feet below me, the Pasterze Glacier. Of course we retreated precipitately ; but nevertheless, I and the leading guide had been standing for some minutes on a shelf of snow which the wind had drifted against the smooth surface of the precipice forming the northern side of the Klein Glockner.

It is wonderful that this shelf, not thicker than three feet where it joined the rock, should have withstood our double weight ; and at the same time it serves to illustrate the incredible force of gales in winter time at high elevations.

The 'saddle' over which we had to pass was a decidedly bad place, and even in summer, when the wire rope that has been fastened across it can be used, every precaution is necessary. Now the rope was invisible, imbedded in ice, in fact, and consequently we were obliged to walk for thirty or forty feet along an edge not broader than nine or ten inches, having on both sides precipices 3,000 and 4,000 feet deep.

To render this feat even more dangerous, the

wind had increased, making it difficult to keep one's equilibrium while balancing oneself across this icy knife-back.

At five minutes to ten o'clock A.M., on January 2, 1875, we five mortals were standing on the top of the Gross Glockner, having successfully accomplished a feat, which, as my guides afterwards hinted to me, they would not repeat for 500 florins each. The men dropped upon their knees, and offered up a short prayer—a proceeding quite unusual with these fearless fellows, showing more than anything else that the dangers we had passed through were exceptionally great.

The cold had abated – 6° R., or 18° F., was quite bearable, but not sufficient to thaw our provisions, which were frozen as hard as stone. The strong schnapps even was in a half-frozen state, and considering the bad nature of the descent and our exhausted condition, we refrained from taking any for fear of evil consequences. The meat, tea, and wine, of which we stood so much in need, had to be returned untasted into our spacious ' Rücksäcke.'

My card, with the date of the ascent and the names of the four intrepid guides scrawled as legibly as my stiff fingers and shaking frame allowed, I deposited in the cairn that had been raised by preceding mountaineers. A large flagstaff, lying buried under the ice and drifted snow, was dug out, and, after fixing the remnants of a red flag on it, was stuck into a deep hole made by means of our sharp-pronged Alpenstöcke.

The view was magnificent beyond description. The sky was of a dark, dead blue, and the air so clear that we could make out peaks never yet seen from the Gross Glockner.

The Ortler and the Bernina group, invisible in summer from this height, were quite distinct, and seemed hardly further off than the Marmolatta peak (in the Dolomites) in summer.

Far beyond the Bernina we perceived rows of glittering rose-tinted giant peaks, though of course the great distance made it impossible to determine their names.

We remained about thirty-five minutes on our elevated post, and then, waving our hats and shouting one simultaneous ' Jodler' as a last greeting to the flag fluttering in the wind, we turned our backs on that well-known cairn, 13,000 feet over the level of the sea.

I had noticed by means of my telescope groups of people standing in front of the Heiligen Blut Church, lying, as it were, at our very feet, and needing but one gigantic leap of some eight or nine thousand feet to reach it. What their feelings were on seeing our flag none but a jealously inclined mountaineer can imagine. Three ˉconsecutive winters had they tried to vanquish the Gross Glockner ; and though they once got as far as the slopes leading to the Klein Glockner, they had on every occasion failed to reach the spot we were just about leaving.

These attempts, I may add, had been made in winters when a much smaller quantity of snow made high elevations less inaccessible. As we looked down the terribly steep slopes, which were one mass of ice, it seemed impossible, unprovided as we were with any instrument to cut proper steps, or to anchor ourselves effectually if one of us slipped, to get down in safety. 'One slip and we are killed,' were the words with which dauntless Peter took the lead down that icy incline.

With the greatest caution, and making use of our crampons, which latter were of the most vital service, we managed to reach the 'Adlersruhe.' From that point to Kals we met with nothing extra-ordinary, excepting one avalanche. It seems strange that in ascending in the cold night we had seen three of them, while on our return in daytime, with a bright sun shining, we only came across one.

So eager were we to reach Kals and announce our success, that our descent from the 'Adlersruhe' was accomplished in double-quick time, the evening-prayer bell (4 o'clock) ringing in our victorious return to Kals. Our flag had been seen, and a large crowd of inhabitants came to meet us and proffer us their congratulations.

A fast of nearly eighteen hours, and great bodily exertions, had left us famishing. Our attacks on food of every sort were closely watched and admired by a crowded audience in the Glockner Wirth's cosy parlour.

LONDON : PRINTED BY
SPOTTISWOODE AND CO., NEW-STREET SQUARE
AND PARLIAMENT STREET

www.ingramcontent.com/pod-product-compliance
Lightning Source LLC
Chambersburg PA
CBHW021502210326
41599CB00012B/1108